Homing the Machine in Architecture

Homing the Machine in Architecture is a series of conversations on the ways designers, practitioners, historians, and theorists orient themselves within the world of architectural digital fabrication. To "home" a digital fabrication machine is to send it back to its origin point—a point that can be specified by the fabricator in advance of the fabrication process or by the defaults that are pre-programmed into the machine. The homing process is necessary and productive since it determines the physical point at which the machine (and the maker) begin making—every time that architectural designers begin to digitally fabricate something new, they first need to home the machine. This book gathers first- and second-hand accounts of the origins of individual "digi-fab" practices from the emergence of advanced prototyping tools to the contemporary moment. It features interviews, essays, and case studies organized around three questions: What are the possible histories of digital fabrication in architecture? How do designers orient themselves in this emergent discipline? What conceptual original points do architectural designers return to when they home their machines?

The discourse that emerges from this collection aims to reach practicing architects using digital fabrication, as well as upper-level students and academics of digital architecture, architectural theory, and architectural history.

Galo Canizares is a designer, writer, and educator. His work blends absurdity, computation, world-making, simulation, and parafiction to address issues in technology and the built environment. He is currently researching the sociotechnical networks of relations between design's softwarization and the architectural imagination. Galo was the recipient of the 2016–2017 LeFevre '29 Emerging Practitioner Fellowship, and in 2018 was awarded the Christos Yessios Visiting Professorship at The Ohio State University. His collaborative architectural practice, office ca, won the 2018 Ragdale Ring competition. He is currently Assistant Professor of architecture at the University of Kentucky College of Design where he directs the Critical Software Lab.

Zach Cohen is a designer, researcher, and educator. His research and teaching examine the ways in which architects can use digital fabrication technologies to reimagine both the immaterial and physical labor of architectural design. From 2019 to 2021, Zach was the Christos Yessios Visiting Assistant Professor at the Knowlton School of Architecture at The Ohio State University. Prior to his time at Knowlton, he was a Research Lead at the Self-Assembly Lab at the Massachusetts Institute of Technology. Zach is currently an Adjunct Assistant Professor at The Cooper Union, where he teaches courses centered on what he calls "theoretical digital fabrication." He is also the co-founder and partner of the Brooklyn-based architectural design practice, common craft, which presently has a variety of commercial and residential construction projects in and around New York City.

Homing the Machine in Architecture

Edited by Galo Canizares and Zach Cohen

Routledge
Taylor & Francis Group
LONDON AND NEW YORK

Cover by Galo Canizares and Zach Cohen

First published 2024
by Routledge
4 Park Square, Milton Park, Abingdon, Oxon OX14 4RN

and by Routledge
605 Third Avenue, New York, NY 10158

Routledge is an imprint of the Taylor & Francis Group, an informa business

British Library Cataloguing-in-Publication Data
A catalogue record for this book is available from the British Library

Library of Congress Cataloging-in-Publication Data
Names: Canizares, Galo, editor. | Cohen, Zach, editor.
Title: Homing the machine in architecture / edited by Galo Canizares and Zach Cohen.
Description: Abingdon, Oxon: Routledge, 2024. | Includes bibliographical references and index. |
Identifiers: LCCN 2023036928 | ISBN 9781032283708 (hardback) | ISBN 9781032283654 (paperback) |
ISBN 9781003296522 (ebook)
Subjects: LCSH: Architectural design—Methodology. | Architecture—Technological innovations. |
Building—Technological innovations.
Classification: LCC NA2750 .H648 2024 | DDC 720.285—dc23/eng/20231115
LC record available at https://lccn.loc.gov/2023036928

ISBN: 978-1-032-28370-8 (hbk)
ISBN: 978-1-032-28365-4 (pbk)
ISBN: 978-1-003-29652-2 (ebk)

DOI: 10.4324/9781003296522

Typeset in Univers
by codeMantra

Contents

Figures

Contributors

Dr. Ehsan Baharlou is an Assistant Professor of Architecture, Advanced Technologies, in the School of Architecture at UVA. His work focuses on integrating design computation and the process of materialization. His research aims to bridge the gap and mediate the cyber-physical interactions between complex forms and advanced manufacturing tools, shifting from a paradigm of abstracted computational design toward an integration of both physical (fabrication and production) and digital investigations. He completed his postdoctoral associate '18 at MIT (Massachusetts Institute of Technology) in the Department of Architecture, in Prof. Mark Goulthorpe's group (Composite Architectures). His post-doctoral research concentrated on the automated production of composite housing and the development of the CAD/CAM software customization in composite architectures. He earned his PhD '17 under the supervision of Prof. Achim Menges at ICD (Institute for Computational Design and Construction), University of Stuttgart. His doctoral research focused on the integration of fabrication and construction constraints into a computational model for the realization of informed form generation.

Nathan Barnes is the Computational Lead at A. Zahner Co. He has a background in architectural fabrication through programmatic mass customization. He is passionate about encapsulating solutions to problems through technological applications. Additionally, he has a passion for fostering an inquisitive culture around him. Nathan holds a Master's degree in architecture from UTA with an emphasis on digital fabrication. His current research is centered around developing software and hardware improvements to nudge forward fabrication capabilities.

Dr. Paolo Cascone was born in Italy and grew up between West Indies and East Africa. He is an AA trained architect with a PhD at the intersection of environmental engineering and sustainable architecture. Paolo is the founding director of CODESIGNLAB (www.codesignlab.org) as well as Senior Lecturer in Architecture and Environmental Design at the University of Westminster in London. He is the scientific director of the African Fabbers project which has been recently displayed at the last Venice Biennale of Architecture (May 2021). Paolo's work on ecological design and performative architecture has been exhibited and published widely.

Mollie Claypool is an architecture theorist and educator focused on issues of social justice highlighted by increasing automation in architecture and design production. She is Associate Professor in Architecture at The Bartlett School of Architecture, UCL, where she is Co-Director of Automated Architecture (AUAR) Labs and History & Theory Coordinator in MArch Architectural Design. Mollie is Co-Founder and CEO of the AUAR Labs spinout AUAR Ltd, a technology company working on distributed automated housing production. Mollie is co-author of the book *Robotic Building: Architecture in the Age of Automation* (Detail Edition, 2019) and author of the SPACE10 report "The Digital in Architecture: Then, Now and in the Future" (2019).

James Coleman is co-founder of SumPoint. He brings a unique mix of design, engineering, construction, and digital fabrication expertise. His work focuses on the comprehensive process of delivering extraordinary architectural projects; leveraging parametric design-to-fabrication-to-installation workflows. Previously, James acted as Vice President - Innovation R&D at Zahner, a company that specializes in computational fabrication at an architectural scale. As a part of the Executive Team, James led research initiatives and provided strategic support on large scale and complex projects. He has worked as a Design Engineer with award winning architecture firms around the world and as a Product Development Engineer at Ford Motor Company James holds a Master of Architecture and Master of Science in Mechanical Engineering from the Massachusetts Institute of Technology and a Bachelor of Environmental Design from the University of Colorado-Boulder.

Shelby Doyle, AIA is a registered architect and Associate Professor of Architecture where she is the Stan G. Thurston Professor of Design Build at Iowa State University College of Design, co-founder of the ISU Computation & Construction Lab (CCL), and director of the ISU Architectural Robotics Lab (ARL). The CCL and ARL the result of Doyle's ISU Presidential Impact Hire to rethink digital fabrication and design-build. The CCL works to connect developments in computation to the challenges of construction: through teaching, research, and outreach. The central hypothesis of CCL and Doyle's work is that computation in architecture is a material, pedagogical, and social project; computation is both informed by and productive of architectural cultures. This hypothesis is explored, through the fabrication of built projects and materialized in computational practices. The CCL is invested in questioning the role of education and pedagogy in replicating existing technological inequities, and in pursuing the potential for technology in architecture as a space of, and for, gender equity. Doyle is currently a Mellon Research Fellow for the Canadian Centre for Architecture's The Digital Now: Architecture and Intersectionality, the 2023 President of ACADIA—the Association for Computer Aided Design in Architecture—and a 2022–2025 At-Large-Director for the Association of Collegiate Schools of Architecture (ACSA).

Dr. Behnaz Farahi is a designer, creative technologist, and critical maker. Trained as an architect, she explores how to foster an empathetic relationship between the human body and the space around it through the implementation of emerging technologies. Her goal is to enhance the interaction between human beings and the built environment by following morphological, and behavioral principles inspired by natural systems. Her work addresses critical issues such as feminism, emotion, bodily perception, and social interaction. She specializes in computational design, interactive technologies, additive manufacturing, and digital fabrication technologies. Farahi has won several awards including the Cooper Hewitt Smithsonian Design Museum Digital Design Award, Fast Company Innovation By Design Linda Tischler Award, World Technology Award (WTN) and is the recipient of the BASA and Madworkshop Grants and the Rock Hudson Fellowship. Her work is in the permanent collection of the Museum of Science and Industry in Chicago. Her work has been exhibited internationally at Ars Electronica, Linz and Context Art Miami, SIGGRAPH, La Piscine Museum in France and A+D Architecture and Design Museum in LA and has been featured in several magazines and online websites including *WIRED*, BBC, CNN, *The Guardian*, *Frame Magazine*, and many more. Farahi has worked with leading firms such as Adidas, Autodesk, Fuksas Studio, and 3DSystems/will-i-am. She has also collaborated with Professor Behrokh Khoshnevis on two NASA-funded research projects developing a robotic fabrication technology to 3D print structures on the Moon and Mars. She has been an Artist in Residence at Autodesk Pier 9. She has received her PhD in Interdisciplinary Media Arts and Practice at the USC School of Cinematic Arts. She also holds a Bachelors and two Masters degrees in Architecture. Currently, she is an Assistant Professor at the Department of Design, California State University, Long Beach.

Dr. Madeline Gannon is a multidisciplinary designer blending techniques in art, design, computer science, and robotics to forge new futures for human-robot relations. Also known as "The Robot Whisperer," Gannon specializes in convincing robots to do things they were never intended to do: from transforming giant industrial robots into living, breathing mechanical creatures, to taming hordes of autonomous machines to behave like a pack of animals. Gannon believes that technology is a cultural medium, and tunes her work to engage communities across science and society. Her works have been exhibited at international cultural institutions, published at academic conferences, and profiled at global media outlets, such as the BBC, *The Guardian*, FT, the Science Channel, *WIRED*, FastCompany, Dezeen, and The Verge. Dr. Gannon is a World Economic Forum Cultural Leader, a former Robotics & AI Researcher at NVIDIA and a former artist in residence at ETH Zurich, Autodesk Pier 9, and the Carnegie Mellon STUDIO for Creative Inquiry. She is known as one of the 'Top 10 Women in Robotics Industry' and 'World's 50 Most Renowned Women in Robotics' according to Analytics Insight. Gannon holds an M.Arch from Florida International University, and a PhD in Computational Design from Carnegie Mellon University.

Pablo C. Herrera develops workshops and lectures incorporating emerging digital technologies in Latin America, and, from that point, the use of self-learning techniques in order to enhance local traditions and patterns, according to culture and reality. He has coordinated and researched space and shape problems using algorithms and programming, establishing since 2006 the implementation of these techniques in many Latin American universities with multidisciplinary post graduate students from USA and Europe. His research and workshops have been presented in several cities such as Santiago de Chile (2006–2007), Mexico (2007), Lima (2007, 2011), Havana (2008), Medellin (2008), Sao Paulo (2009), Bogota (2009–2010), and Valparaíso (2009, 2011). Since 2007, he has published refereed papers, journal articles, some of which won best presentation's awards. He is two-term vice president and an elected board member of SIGraDi (Sociedad Iberoamericana de Gráfica Digital) and elected editorial board member of IJAC (International Journal of Architectural Computing). He is also peer-reviewer at CAADRIA, eCAADe, ICDHS, CAAD Futures and SIGraDi. He was one of the six Latin Americans invited to exhibit and publish the results of their academic work at the III Architecture Biennial Beijing (2008) and at the IV Architecture Biennial Beijing: Machinic Processes (2010). In the last ten years, he has provided independent consulting in various Latin American universities and companies (including the FABLab Lima), seeking to manage and implement digital technologies in design and manufacture. He has been Keynote Speaker in several conferences, including the XII Congress of SIGraDi (2008), at the XX Congress of ACFA (Association of Schools of Architecture, Bogotá, 2009), at the Politics of Fabrication Symposium (Sponsored by the Architectural Association, Valparaiso, 2011) and since 2011 is Workshop Coordinator at SIGraDi.

Francesca Hughes, an architect by training, is an internationally renowned architectural theorist and educator. Having taught across both studio and history & theory at the Bartlett School of Architecture, UCL, and then the Architectural Association in London for more than 25 years, she was a Professor of Architecture and Head of School at UTS, Sydney, from 2018 to 2020. Given her extensive knowledge of different architectural education cultures and models, Francesca has served as visiting professor and external examiner at numerous schools in the UK, US, Asia, and Europe. *The Architect: Reconstructing her Practice* (MIT Press, 1996), an early focus on feminist criticism and architecture, sits at the foundation of her larger body of published work which examines architecture's thorny relations to technology. To this end, her written and design research spans a critique of reductive superficiality in architecture's use of computation in Drawings that Count (AA Publications, 2013), an examination of architecture's relations to precision and material error in The Architecture of Error: Matter, Measure and the Misadventures of Precision (MIT Press, 2014), and an exploration of architecture's relations to systems of measure and prediction in Architectures of Prediction (ARQ docs, 2019). Her current project, The Architect's Computer: An Indiscrete History of the Universal Discrete Machine, is an histoire longue durée of the discipline's relations to the constituents of modern and ancient computation: the algorithm as calculator of new "truths"; the cognitive interface and its universal languages; memory storage and retrieval; deletion and forgetting; and the window. Having collaborated with historians of science, archaeologists, filmmakers, robotic landscapers, and performance artists, Francesca remains intent on expanding architecture's perimeter and an ardent advocate of all things rigorously multi-disciplinary. Francesca has lectured internationally and acted as keynote to numerous conferences. Her written and built work have been published by AA Files, AD, ANY, AR, A/R/P/A (Columbia GSAP), Art Forum, Columbia University Books on Architecture, e-flux Architecture, Harvard Design Magazine, e-flux, Merrell, Park Books, Random House (Monacelli), Routledge, University of Minnesota Press and Wiley and exhibited in the UK and abroad.

Mark Jarzombek is a Professor of the History and Theory of Architecture at MIT. Jarzombek works on a wide range of topics—both historical and theoretical. He is one of the country's leading advocates for global history and has published several books and articles on that topic, including the ground-breaking textbook entitled *A Global History of Architecture* (Wiley Press, 2006) with co-author Vikramāditya Prakash and with the noted illustrator Francis D.K. Ching. He is the sole author of *Architecture of First Societies: A Global Perspective* (Wiley Press, 2013), which is a sensitive synthesis of first society architecture through time and includes custom-made drawings, maps, and photographs. The book builds on the latest research in archeological and anthropological knowledge while at the same time challenging some of their received perspectives. Jarzombek also recently published a book that interrogates the digital/global imaginaries that shape our lives. A chapter from that book is available online. *Digital Stockholm Syndrome in the Post-Ontological Age* (University of Minnesota Press, 2016). Jarzombek's ground-breaking work on global architecture history was highlighted by a 2.5 million dollar grant from the Andrew W. Mellon Foundation that Jarzombek received with co-PI, Vikramāditya Prakash (University of Washington, Seattle), to create a new scholarly entity called Global Architecture History Teaching Collaborative (GAHTC). Promoting the development and exchange of teaching materials for architectural history education across the globe, the collaborative provides awards to members and their teams to develop new lecture material from global perspectives.

Maddalena Laddaga graduated in architecture at the Bari polytechnic university. She has developed research projects in India and worked in architectural firms between Rome and Berlin. In the recent years, she has developed applied researches in the field of digital fabrication related to architecture. Maddalena is co-director of the CODESIGNLAB since year 2020 and she was research associate of the African Off-grid housing project at the University of Westminster in London. Maddalena is currently PhD candidate at DAStU-Politecnico di Milano Department of architecture and urban studies.

Leslie Lok is a co-principal at HANNAH, an experimental design and research studio. Working across scales, the office has a keen interest in architectural explorations grounded in material expression, digital fabrication, and construction. HANNAH's recent work includes the 3D-printed RRRolling Stones furniture at Socrates Sculpture Park in Long Island City, New York, Ashen Cabin, a robotically fabricated building in Ithaca, New York; and House of Cores, the first two-story 3D printed building in the U.S. HANNAH is the recipient of the 2020 Architectural League Prize, was named one of *Architect Magazine*'s Next Progressives in 2018 and won ArchDaily's Best New Practice Award in 2021. Lok is also an assistant professor at Cornell University AAP where she directs the Rural-Urban Building Innovation (RUBI) Laboratory, which investigates new building methods for the design of adaptable buildings and housing in rural-urban contexts. Leveraging computational protocols, digital fabrications, and technologies like augmented reality (AR) and mixed reality (MR) tools, RUBI Lab explores how local building practices can incorporate both traditional and innovative building materials and fabrication techniques to accommodate changing environments and lifestyles across the urban fringe and rural contexts. Lok received her Master of Architecture degree at Massachusetts Institute of Technology.

Zachary Mollica is an architect, educator, and maker known for his work with non-linear parts of trees and other unusual bits of wood. He combines digital techniques, like 3D scanning, with traditional craft and material knowledge in pursuit of better ways to build. In 2022, Mollica became the first Emerging Architect Fellow at the John H. Daniels Faculty of Architecture, Landscape, and Design. With collaborators and graduate students, his research pursues the development of design and technical tools intended to spur a re-diversification of wood building methods. By close observation of forests and the inherent properties of trees and wood, Mollica's work demonstrates new processes that enable forest products of little economic value to be converted into high-value building components with minimal energy. The approach is most fully represented by the Tree Fork Truss project, a large spanning structure that Mollica led the design and assembly of 2015. Before his current position, Mollica was Warden of the AA School's forest campus, Hooke Park, a unique educational facility in Dorset, United Kingdom. There, he was also a supervisor for AA Design + Make students (MARCH/MSC) and was the founding director of the AA Wood Lab.

Vernelle A. A. Noel, PhD is a computational design scholar, architect, artist, and Director of the Situated Computation + Design Lab. Currently, she is the Lucian and Rita Caste Assistant Professor in Architecture and Urban Design at the Carnegie Mellon University School of Architecture. She investigates traditional and digital practices, interdisciplinary creativity, and their intersections with society. Dr. Noel builds new expressions, tools, and methodologies to explore social, cultural, and political aspects of making, computational design, and emerging technology for new social and technical reconfigurations of design practice, pedagogy, and publics. Her work is thoroughly interdisciplinary with training in design computation; architecture; science, technology, and society (sts) studies; media arts, and sciences. Dr. Noel's work has been supported by the Graham Foundation for Advanced Studies in the Fine Arts, the Mozilla Foundation, and ideas2innovation (i2i), among others. Her TEDx Talk is titled, "The Power of Making: Craft, Computation, and Carnival." She is a recipient of the 2021 DigitalFUTURES Young Award for exceptional research and scholarship in the field of critical computational design. Vernelle has been a Visiting Professor at the University of Stuttgart, faculty at the University of Florida, instructor and Visiting Fellow at Penn State University, The

Massachusetts Institute of Technology (MIT), and the Singapore University of Technology & Design. She was a Keynote Speaker at ACADIA 2020 and has been featured on Madame Architect. Vernelle holds a PhD in Architecture (Design Computing) from the Pennsylvania State University; a Masters of Science in Architecture Studies from MIT (Design Computation); a Bachelor of Architecture from Howard University; and a Diploma in Civil Engineering from the John S. Donaldson Technical Institute (University of Trinidad & Tobago). She has practiced as an architect in the USA, India, and Trinidad & Tobago and is the founder, creator, and editor of Architecture Caribbean, an online platform that showcased and promoted design by Caribbean nationals. Noel has served on the Board of ACADIA since 2021.

Prof. Stefana Parascho is Tenure Track Assistant Professor in the Laboratory for Creative Computation at EPF Lausanne. Her interest in geometry and computational design has developed in her studies in Bucharest, Stuttgart and at ETH Zurich where she was a PhD researcher in the NCCR DFAB from 2014 to 2018. She has also worked professionally in engineering, and in computational production. As a PhD researcher at the Chair of Architecture and Digital Fabrication at ETH Zurich, her research focused on complex robotic assembly and path planning. In 2019, she became an Assistant Professor at the University of Princeton. In 2022, she returned to Switzerland and launched the Laboratory for Creative Computation at EPF Lausanne.

Marshall Prado is an Assistant Professor of Design and Structural Technology at the University of Tennessee. He holds a Bachelor of Architecture and a Bachelor of Environmental Design from North Carolina State University and advanced degrees of a Master of Architecture and a Master of Design Studies in Technology from the Harvard University Graduate School of Design. Marshall has previously taught at the University of Stuttgart, in the Institute for Computational Design and Construction, the University of Hawaii, and has been an invited studio critic at the University of Virginia, Carnegie Mellon University, and the University of Michigan. He has led several workshops on digital design and fabrication techniques. His current research interests include integrated computational design, robotic fabrication, additive manufacturing, and lightweight fiber composite systems for architectural applications.

Michael Szivos is the founder of SOFTlab, a New York City based design studio that combines research, technology, and ideas to craft work that spans multiple mediums and scales. In 2012, SOFTlab was awarded the Architectural League Prize for Young Architects & Designers, and in 2010 the studio was selected, along with seven other young studios, for the New Practices New York award by the AIA Chapter of New York. He also teaches at Pratt Institute and Yale University.

Skylar Tibbits is a designer and computer scientist whose research focuses on developing self-assembly and programmable materials within the built environment. Tibbits is the founder and co-director of the Self-Assembly Lab at MIT, and Associate Professor of Design Research in the Department of Architecture. He is also the director of undergraduate programs in the Department of Architecture. Tibbits has a professional degree in architecture and a minor in experimental computation from Philadelphia University, and a masters in design computation and masters computer science from MIT. He has worked at a number of design offices including Zaha Hadid Architects, Asymptote Architecture and Point b Design. He has designed and built large-scale installations and exhibited in galleries around the world, including the Centre Pompidou, Philadelphia Museum of Art, Cooper Hewitt Smithsonian Design Museum, Victoria and Albert Museum and various others. He is the author of *Things Fall Together: A Guide to the New Materials Revolution* (Princeton University Press, 2021), *Self-Assembly Lab: Experiments in Programming Matter* (Routledge, 2016), *Active Matter* (MIT Press, 2017), co-editor of *Being Material* (MIT Press, 2019), and the Editor-In-Chief of the journal *3D Printing and Additive Manufacturing*. Awards include LinkedIn's Next Wave Award for Top Professionals under 35 (2016), R&D Innovator of the Year (2015), National Geographic Emerging Explorer (2015), an Inaugural WIRED Fellowship (2014), the Architectural League Prize (2013), Ars Electronica Next Idea Award (2013), TED Senior Fellow (2012) and 2008 he was named a Revolutionary Mind by SEED magazine.

Theodora Vardouli, PhD is an Assistant Professor at the Peter Guo-hua Fu School of Architecture at McGill University, where she researches histories, cultural meanings, and operational implications of algorithmic techniques for architecture. Vardouli has published articles in *Architectural Research Quarterly, Design Studies, IEEE Annals of the History of Computing, Nexus Network Journal, Perspective, Leonardo,* and chapters in several edited collections. She is the author of the forthcoming book *Graph Vision: Digital Architecture's Skeletons* (MIT Press, 2024), co-editor of *Computer Architectures: Constructing the Common Ground* (with Olga Touloumi, Routledge, 2020) and co-author and co-editor of *Designing the Computational Image, Imagining Computational Design* (with Daniel Cardoso Llach, Applied Research and Design Publishing, 2023). Vardouli also co-curated with Cardoso Llach the exhibition *Vers un imaginaire numérique* (Montreal 2021). Vardouli's scholarship has been recognized with the 2022 Best Paper Award by the Association for Computer Aided Design in Architecture (ACADIA) and the 2022 Mahoney Prize, for an article co-authored with David Theodore, by the Special Interest Group for Computing Information and Society (SIGCIS). Vardouli is an External Examiner for two graduate programs at the Bartlett School of Architecture, University College London and the Communications Editor for the journal *Technology, Architecture + Design* (*TAD*).

Maria Yablonina is an architect, researcher, and artist working in the field of computational design and digital fabrication. Her work lies at the intersection of architecture and robotics, producing spaces and robotic systems that can construct themselves and change in real-time. Such architectural productions include the development of hardware and software solutions, as well as complementing architectural and material systems in order to offer new design spaces. Maria's practice focuses on designing machines that make architecture—a practice that she broadly describes as Designing [with] Machines (D[w]M). D[w]M aims to investigate and establish design methodologies that consider robotic hardware development as part of the overall design process and its output. Through this work, Maria argues for a design practice that moves beyond the design of objects towards the design of technologies and processes that enable new ways of both creating and interacting with architectural spaces. Currently Maria is an Assistant Professor at the Daniels Faculty of Architecture, Landscape and Design at the University of Toronto.

Sasa Zivkovic is a co-principal at HANNAH, an experimental design and research studio. Working across scales, the office has a keen interest in architectural explorations grounded in material expression, digital fabrication, and construction. HANNAH's recent work includes the 3D-printed RRRolling Stones furniture at Socrates Sculpture Park in Long Island City, New York, Ashen Cabin, a robotically fabricated building in Ithaca, New York; and House of Cores, the first two-story 3D printed building in the U.S. HANNAH is the recipient of the 2020 Architectural League Prize, was named one of *Architect Magazine*'s Next Progressives in 2018 and won ArchDaily's Best New Practice Award in 2021. Zivkovic is also an assistant professor at Cornell University AAP where he directs the Robotic Construction Laboratory (RCL), an interdisciplinary research group that develops and implements novel robotic construction technology. Interdisciplinary in nature, the work integrates cutting-edge materials, advanced fabrication, mechanical design, architectural computation, structural optimization, and sustainable construction. His academic research has been published by FABRICATE, Rob|Arch, ACADIA, ACSA, and RILEM Digital Concrete, among others. Zivkovic received his Master of Architecture degree at Massachusetts Institute of Technology.

Acknowledgments

The idea for this book originated at the "Homing the Machine" symposium at The Ohio State University's Knowlton School of Architecture in January 2020. For this event, we invited a group of 13 architecture-oriented digital fabrication researchers and practitioners to reflect on this subdiscipline's comparatively short, yet full, history in architectural research and practice. We asked this group of colleagues to discuss how they orient their work relative not only to digital fabrication's recent past, but also to architectural discourse in general. We are incredibly thankful to the original participants, some of whom also contributed to this volume: Skylar Tibbits, Marshall Prado, Maria Yablonina, Molly Wright Steenson, Tsz Yan Ng, Joshua Stein, Shelby Doyle, Justin Diles, Michael Szivos, James Coleman, Sasa Zivkovic, Mariana Ibanez, and David Benjamin. The symposium and its related research were funded by the Knowlton School of Architecture at Ohio State and the Christos Yessios Visiting Professorship.

We are immensely thankful for the opportunity to collaborate with all the contributors included here. Editing a book is so much more than collecting and compiling different essays. Each piece is the product of an extensive dialogue between us and the authors. We are grateful not just for their expertise, but also for their generosity, insights, and extensive time commitments that this project required. We would also like to thank the editorial team at Routledge and colleagues at our respective institutions. This book would not have been possible without such a supportive network of people.

Homing

Galo Canizares and Zach Cohen

Homing the Machine

To "home" a digital fabrication machine is to determine its origin point: the point in space at which the machine begins making. Homing is necessarily the first step in every digital fabrication process since it provides the machine with addressable spatial coordinates that will serve as a reference for all subsequent coordinates that it will visit along its "tool path." Without these origins, digital fabrication machines would quickly get lost.

When architects digitally fabricate something, they either specify their own origin points or use the default origins that are pre-programmed into machines by their manufacturers. If an architect specifies a machine's origin point, they often do so in consideration of the thing they intend to make, for example, its size or fragility. Thus, each time an architect decides to digitally fabricate something different, they may also decide to change the digital fabrication machine's origin. In this way, origin points demonstrate how *what* architects decide to digitally fabricate is intertwined with *how* they decide to digitally fabricate, suggesting that *physical* origins always in some way follow *conceptual* ones.

Since digital fabrication's introduction to the field of architecture, architects have conceptualized various things for digital fabrication machines to make. This research has produced a variety of approaches and ideologies, as well as many different origin points. And yet, today, the discourse of digital fabrication in architecture generally remains centered on the technically oriented rhetoric inherent to digital fabrication's industrial

DOI: 10.4324/9781003296522-1

origins—buzzwords like *streamlining* and *optimization* still abound—and on tools rather than values. The friction between the persistently technical nature of this type of research and the undeniably conceptual nature of architectural design has made it difficult to locate digital fabrication's home, or homes, within the greater discipline of architecture. But given the amount of subjectivity involved in acts of digital fabrication, from homing the machine to realizing digitally fabricated assemblies, we believe it is time to look into the decision-making processes behind the more often discussed technical details; in other words, we want to ask: what are the conceptual origins from which our physical origins emerge? And how have these conceptual origins informed the kinds of things architects and designers have decided to digitally fabricate?

Our use of "origin" here operates in two ways. First, as described above, it provides a methodology for discussing orientation in the discipline of architecture. Architecture has many subdisciplines, of which digital fabrication is just one, and it is easy to get lost in conceptual discussions of disciplinary lineages, projects, and themes. Thinking about the origin as a starting point allows us to trace vectors, paths, and trajectories related to this particular subdiscipline. Second, we are looking to reflect on the history of digital fabrication as an evolving, multiplicitous subject. We posit that this history is made up largely of personal microhistories and intimate experiences with machines and technologies that do not necessarily constitute a linear concrete chronology. Thus, the best way to chronicle the history of digital fabrication in architecture may be to look at personal "origin stories" concerning individual practices and approaches to emergent fabrication technologies. In this book, we aim to take stock of these origins in order to both situate digital fabrication's present role in architecture and begin to plot its path moving forward.

Multiple Origins

Digital fabrication has a comparatively short, yet full, history as a subdiscipline in architectural research and practice. In conducting research for this book, we asked a number of our colleagues to describe how they orient their work relative not only to digital fabrication's recent past but also to architectural discourse in general. One finding that emerged from these queries was that each person had a different account of when digital fabrication actually *became* a subdiscipline in architecture; in other words, the introduction of digital fabrication to architecture was yet another origin that, perhaps necessarily, varied from architect to architect. For example, some participants contended that the unprecedentedly complex construction of Frank Gehry's Museum of Pop Culture (Figure 1.1), completed in 2000, marked digital fabrication's first architectural appearance; others suggested that Bill Massie's winning entry to the 2002 P.S.1 Young Architects Competition, "Playa Urbana," was when architects first gleaned the possibilities of CNC (computer numerically controlled) fabrication (Figure 1.2); and still others argued that the *architectural* variety of digital fabrication (which is seemingly distinct from the industrial or scientific varieties) was initiated when Gramazio Kohler Research first published their robotic brick-stacking in 2006 (Figure 1.3).

Figure 1.1 Frank Gehry's Museum of Pop Culture (originally the Experience Music Project Museum) with complex curved cladding designed by A. Zahner.

Figure 1.2 Playa Urbana/Urban Beach by Bill Massie. Installed at MoMA PS1 as part of the 2002 Young Architects Program.

This apparent mutability of historical beginnings suggests that the origins of digital fabrication in architecture depend crucially on who is telling the story. As such, its historiography could be regarded more as a loose collection of interrelated episodes rather

Figure 1.3 The Endless Wall by Gramazio Kohler Research, ETH Zurich, 2011.

than a singular consensual genealogy of events and projects. Unlike the grand narratives of modernism and postmodernism, for instance, digital fabrication in architecture was driven primarily by small technological experiments absent from dominating manifestos. This praxis-oriented perspective resulted in a variety of concurrent yet incongruous projects that can each be argued to have catalyzed the subdiscipline of digital fabrication.

As the architectural historian Anthony Vidler has noted, architectural history is a "determining, unfolding force, capable of articulating questions of the past, present, and future of architecture, as well as a belief in some form of sociocultural zeitgeist."[1] Seen in this light, the history of digital fabrication in architecture can be a critical lens through which to view contemporary architectural practices. But, as Vidler also states, architectural history is mutable, particularly in cases where historians and critics seek to justify contemporary work through historical alibis. In the case of modernism, for example, architectural historians often developed

> their genealogies from moments in the past that seemed to them starting points that would justify the specific contemporary practices they supported or admired [...] Each of these different histories imagined modernism in a form deeply complicit with the 'origin' it proposed.

These varied histories were often entwined in a feedback loop between projects and practices that ostensibly fit neatly into a singular critical perspective. For Vidler, the task of the architectural historiographer is thus to unpack all of these histories and reveal both conflict and consensus among them.

Though digital fabrication is a relatively new subdiscipline in architecture, there have been several recent attempts to recount its development and influence on architectural design. These histories of digital fabrication follow Vidler's notion of complicity in their argumentation of origins, teasing out individual threads that tidily tie projects and abstract concepts together. For example, in his book, *Builders of the Vision: Software and the Imagination of Design*, architect Daniel Cardoso Llach distinguishes the history of digital fabrication in architecture as principally a history of hardware. According to Llach, digital fabrication, specifically CNC-machining, was initialized by MIT Professor Douglas T. Ross' invention of the "Automatically Programmed Tool" (APT) in 1956. In his account of the earliest computer-controlled machining tools, Llach notes that there were also many "pseudo-digital" predecessors to the APT, for example, John T. Parson's "Numerically-Controlled Machine" (1952) and, much earlier, Joseph Jacquard's loom (1801). It becomes increasingly evident that defining one singular starting point is nuanced and subjective. Yet, what Llach makes abundantly clear is that, no matter the precise date, it was hardware that drove the subdiscipline's development.

Other accounts place digital fabrication's origins in the democratization of design via desktop computing—that is, when architects in offices and schools began to design complex forms with computer software. Greg Lynn is one architect who has championed such a software-centric narrative. In *Archaeology of the Digital*, Lynn discusses the software practices of the 1980s at length and suggests that the origins of digital fabrication are inextricably linked to the emergence of digital design. Digital design software enabled architects to conceive and visualize complex forms, but architects then needed a means to translate these virtual forms into material things: enter digital fabrication. In effect, Lynn's history, or "archaeology," begins when architects like Christos Yessios and Frank Gehry were actively pursuing new forms of computer-aided design; in the former's case, developing software from scratch (ARCHIMODOS), in the latter's case, appropriating existing software from other industries (CATIA). In contrast to Llach, the emphasis here is on software over hardware, a notion clearly evident, for example, in one of Gehry Partners' earliest and most famous projects, the Guggenheim Bilbao. As architecture critic Ashley Mendelsohn has written, while the Guggenheim was without question a turning point in architectural innovation, "some of the greatest moments of ingenuity during the building's design and construction were distinctly low-tech." Though designed digitally, Bilbao's fabrication was far from fully computerized, relying heavily on traditional means of manufacturing and assembly, and thus, inadvertently posing the question: at what point did digital design methods necessitate fully digital fabrication methods?

Between these two beginnings, which pit hardware origins against software origins, is a more recent account by Zeynep Çelik Alexander, which argues that neither camp is solely responsible. Instead, Alexander's book, *Design Technics: Archaeologies of Architectural Practice,* suggests that the instrumentalization of both hardware and software in different contexts provides different origin stories. In other words, each architectural technique—modeling, rendering, specifying—has varied relationships to technology and epistemic thought. Therefore, "architecture's epistemic jurisdiction," what architects conceive of as belonging to their discipline, relies on the different hardware and software that have been instrumentalized throughout the discipline's history. From this point of view, CNC-machining, parametric modeling, mixed reality fabrication, etc., all have their own starting points and subsequently have branched off in distinct directions. Alexander's premise also allows for speculation on the social dimension of tools themselves. If each tool is afforded its own history, for instance, then we could frame the history of digital fabrication as a collection of microhistories (following Carlo Ginzburg's concept of microhistory) that brings together the economics and politics of various individual tools such as CNC routers, laser cutters, or 3D printers.[2] This perspective is crucial for any historiography of digital fabrication as it sheds light on how today's growing assemblage of making equipment found its way into the architectural discipline, as well as how digital fabrication has been both driven and influenced by socioeconomic phenomena such as DIY culture and the maker movement.

Rather than attempt to adjudicate the correctness of these various accounts and/or synthesize them into a single, unified genealogy of digital fabrication in architecture, this book is more interested in how such a multiplicity of historical perspectives—or, origins—has contributed to the current varieties of digital fabrication work. Or, to frame the question historiographically: in what ways have digital fabricators fabricated histories that produced or justified their practices? This book explores potential answers to this question by revisiting and reorienting digital fabrication's architectural beginnings; it uses a multi-disciplinary approach not only to tell diverse stories of how digital fabrication became what it is today but also to reveal the techno-social networks of relations between fabricators and the things they fabricate.

Toward the Techno-Social

This book comes at a pivotal moment for architectural practitioners and educators: pressing issues related to social and environmental crises are revealing the often ignored, or subconsciously suppressed, blind spots in the discipline. In this critical, even confrontational, moment, the architectural subdiscipline of digital fabrication cannot withdraw into its familiar territory of technocentric discourse. Our origin points are determined relative to our machines, but they also exist relative to specific workspaces, which, in turn, are shaped by particular social, political, and environmental contexts—the origin point is, by definition, a situated thing. Therefore, in addition to exploring innovative fabrication processes and construction methods, digital fabrication researchers should also begin to investigate questions of identity and accessibility. For instance, who is invited to innovate with digital fabrication technologies? What ideas are excluded in such innovation? How might biases in software and hardware design impact the design of digitally fabricated things? And, how is digital fabrication knowledge disseminated globally? In a broader sense, these questions touch on the social dimension of technological innovation and research—what we call the *techno-social*—a reminder that, as architectural theorist John May suggests, "nothing technical is ever merely technical." We argue that the time has come to reorient digital fabrication's discursive compass toward the techno-social as opposed to the technocentric.

Prioritizing the techno-social dimension can help us, architects, expand the perception of digital fabrication beyond a subject exclusively about the making of things. For instance, running a fabrication laboratory ("fab lab") relies as much on social dynamics as technical expertise. Yet, from a sociological perspective, there is little doubt that fab labs are often male-dominated or highly masculinized spaces that, unwittingly or not, reinforce gender stereotypes. We can therefore assume that the fab lab, as a social institution, unquestionably has its own politics that produces subjectivities. In fact, architect Randy Deutsch has even dubbed the subject that engages the fab lab: the superuser. In *Superusers: Design Technology Specialists and the Future of Practice*, Deutsch outlines a series of characteristics that some recent graduates of architecture schools exhibit, specifically their savviness with hardware and software; he concludes that these nascent technology specialists are vital for architecture firms today.[3] Deutsch's analysis of a new kind of "generalist" architect-designer places the person running the machine at the center of the discussion. This focus on people, or perhaps even personality, reveals a need to examine more closely how the social interactions between people in fab labs affect the things they fabricate.

The contexts in which fab labs operate also warrant further scrutiny. There is no question that, for a variety of socioeconomic reasons, digital fabrication discourse tends to be dominated by the Global North. Yet, the international network of fab labs supported by the Fab Foundation suggests that the Global South also contributes greatly to this field. The global dissemination of knowledge through academic migration and online resource sharing, for example, has produced an international culture of makers and artisans that blend regional craft with high-tech workflows. These types of subject-oriented dialogues bring to the fore some of the sociological perceptions of digital fabrication practices and pedagogy that have been overshadowed by an emphasis on novel objects and methodologies. Our aim in bringing tangential perspectives, such as May's or Deutsch's, together is to stimulate a discussion surrounding digital fabrication that can capture its multiplicity and far-reaching social effects.

To fully embrace what we're calling the techno-social, it is also necessary to adopt a multi-disciplinary approach to categorization and historical analysis. As architectural researchers, educators, and practitioners ourselves, we have found it difficult to find critical discourses of technology that cross disciplinary boundaries or that propose hybridity as a mode of analysis. At many digital fabrication conferences, for example, it is common to have projects and papers divided into fixed technological categories such as "Augmented Reality Fabrication," "3D Printing," and "Robotic Fabrication." While this may be necessary from an organizational standpoint, these modes of grouping researchers and practitioners not only tend to silo discussions but also often result in highly technical dialogues among so-called experts in a given field. We propose that a loosening of these categorization methods may allow for discovery of unexpected commonalities or perhaps even productive tensions among tangential fields. And so, in this book, we advocate for multiple messy histories over singular neat ones, hybrid categories over fixed classifications, and blurry disciplinary boundaries over hard edges; these criteria drive the structure of the following pages.

Initial Bearings

This book is broken down into three sections, each with an overarching theme: The Technological, The Material, and The Social. These categories are by no means exhaustive or fixed—digital fabrication work extends well beyond these three concepts. Further, as we alluded to above, our employment of such a simple taxonomy is also not meant to suggest that any of the work presented here can or should be so easily reduced.

Of the book's three themes, The Technological is likely the one that will be most familiar to those with previous exposure to digital fabrication discourse. Therefore, this seemed to us the natural point from which to set out our discussion. However, in sharing their personal digital fabrication microhistories, contributors in The Technological section necessarily raise questions and concerns related to The Material and The Social, and even to those sections that the book only implies: The Political, The Cultural, The Environmental, etc. We therefore hope that any expectation the reader may have about which category digital fabrication truly belongs to is quickly dispelled and that they ultimately traverse the book's intentionally loose-knit structure however they see fit. In this way, the book's categories are really meant to perform more like cardinal directions: readers are encouraged to use them to continually orient and reorient themselves as they navigate the diverse perspectives and works that are discussed. In doing so, we imagine that readers will make latent connections and synthesize our initial bearings into hybrid trajectories such as the socio-material or, as we suggested earlier, the techno-social.

Each section comprises case studies and an interview. Between each section is a speculative essay—what we call a Reference Point. This substructure is designed to put technical considerations in conversation with theoretical concerns. We see readers as moderators of this dialogue, simultaneously negotiating the different viewpoints of the book's contributors and channeling questions from worlds outside of digital fabrication. We hope that such heuristic moderation will cause new narratives to emerge.

The first Reference Point in the book lays the groundwork for much of the ensuing discourse. In this essay, the architectural historian Theodora Vardouli historiographically traces digital fabrication's many "homes" relative to technological innovations from the middle of the 20th century to the present day. At each point, Vardouli describes digital fabrication technology not in isolation, i.e., as an "engineering revolution," but instead as a component of larger cultural shifts, such as the maker movement and decolonial computing. In Vardouli's view, homing is both a way of looking back and a program of action for moving forward.

In The Technological section, we look at digital fabrication approaches that originated primarily from experimentation with emergent fabrication tools or techniques. The first case study in this section is by the designer Michael Szivos. Szivos describes how the organic evolution of his experimental design and fabrication practice, SOFTlab, was fueled by the continual discovery of, and subsequent deep dive into, productive technological problems. This notion—that technology can present architects with productive problems—is echoed in the following case study, which is by designer, researcher, and artist Maria Yablonina. In her contribution, Yablonina argues for the architectural and critical value of making non-versatile, task- and site-specific digital fabrication machines, that is, machines that are "stubborn" in their machinic identities. Yablonina suggests that if designers learn to appreciate the idiosyncrasies of such stubborn machines, then we will begin to see them as co-creators, even companions. Yablonina's work thus portends the transition from the purely technological to the more nuanced techno-social. We then turn to the work of Stefana Parascho. Similar to Yablonina, Parascho suggests that interactive and improvisational robotic fabrication can help architects overcome their preoccupation with certain technological tropes. However, whereas Yablonina positions "versatility" as digital fabrication's most detrimental fixation, Parascho proposes that the field's most self-destructive tendency is "optimization." In The Technological interview, we speak with robotic fabrication researcher Marshall Prado, who not only paints a slightly more optimistic picture of robotic fabrication in architecture but also readily admits some of the field's most glaring limitations. Prado shares what it would take for robotic fabrication to work past these limitations and integrate into the mainstream building industry.

In the book's second Reference Point, the architect and theorist Francesca Hughes loops us through competing architectural origin stories and the ways in which they relate to, and potentially derive from, architects' varying understandings of embodiment. Hughes weaves an interplay between two physical centerpoints—the navel and the nozzle—in order to reveal how architects have increasingly encapsulated corporeality, along with the intelligence that goes with it, into machines. She argues that through such transposition, architects have merely displaced the timeless, and distinctly human, task of finding one's center into the machine.

In The Material section, we consider digital fabrication approaches that originated mainly from an interest in the ways in which emergent fabrication tools can invent (or, reinvent) materiality. In the first case study in this section, architects Leslie Lok and Sasa Zivkovic provide an extensive overview of the methods and values embedded within their materially driven digital fabrication practices at their design studio, HANNAH, and their individual research laboratories at Cornell University. In their case study, Lok and Zivkovic echo Prado's assertion that digital fabrication's real-world application is dependent on a revaluation of its aims. However, Lok and Zivkovic further claim that part of the digital fabrication's relegation to academia is that researchers don't sufficiently consider their aims to begin with. Architect and educator Zachary Mollica follows with a provocation that we can and should use digital technologies to rethink our relationships with, and uses of, one of the oldest building materials in the world: wood. In the final material-centered case study, the computational designer Ehsan Baharlou discusses his ongoing efforts to synthesize traditional sustainable construction methods with contemporary ecological principles to put forth a new digital fabrication approach he calls "eco-tectonics." We conclude this section with a dialogue with the founder and co-director of the MIT Self-Assembly Lab, Skylar Tibbits. In this conversation, Tibbits speaks about why embracing failures is critical for our collective cultivation of "material intelligence." Tibbits also repeats the call for a reappraisal of digital fabrication's objectives and suggests, similar to the section's other contributors, that the field's "north star" could be the reduction of architecture's carbon footprint.

In the third Reference Point, the architectural theorist and historian Mark Jarzombek encourages us, as both architects and humans, to look beyond our historical, Western-dominated views of both automation and the body in order to imagine other kinds of robot bodies. Jarzombek's provocation recalls Hughes' disillusionment with the popular Cartesian mind-body dualism that has informed technological practices, both in architecture and beyond.

The book's third section is The Social. In this extended section, we focus on digital fabrication approaches that originated primarily from an interest in and concern for the ways in which emergent fabrication tools can shape new social interactions and cultural phenomena. In The Social's first case study, architectural researchers James Coleman and Nathan Barnes reflect on their work at Zahner, the fabrication company responsible for Gehry Partners' Museum of Pop Culture (Figure 1.3), among many other notable architecture projects. Coleman and Barnes use the practice they developed at Zahner to advocate for developing digital fabrication methodologies in closer collaboration with their primary end users: craftspeople. The next case study by architects Paolo Cascone and Maddalena Laddaga presents a different picture of participatory digital fabrication and craft. In their two endeavors, the CODESIGNLAB and the African Fabbers Project, Cascone and Laddaga strive to work with indigenous African building technologies and local African builders to invent new fabrication techniques and social spaces. These approaches shed light on how fabrication technologies and knowledge have been implemented in the Global South. Along similar lines, architectural researcher and professor Pablo C. Herrera follows with an extensive case study on the evolution of the maker movement and fab labs in Latin America. Herrera reveals how academic migration, government initiatives, and global development contributed to the landscape of digital fabrication in many Latin American countries. For Herrera, the field finds its bearings in the networks of academics and gatherings that promote knowledge sharing. In this section's last case study, architect and professor Shelby Doyle provides insight into her founding of the Computation + Construction Laboratory (CCL) at Iowa State University's College of Design. Doyle also details her development of "computational feminism"—a feminist approach to digital fabrication that informs much of the CCL's research and pedagogy. The case studies of the distinct workspaces presented in this section—industrial, communal, and academic—demonstrate the need for accounts of digital fabrication as a social act between various players, for example, designers and laborers or teachers and students. The Social closes with an interview with computational designer and educator Vernelle A.A. Noel. Through the conversation, Noel lays out the guiding principles of a speculative design practice that she calls "situated computation." In contrast to most digital fabrication approaches, which center on tools and techniques, situated computation stems from people and their regional contexts.

Reorientations

It is admittedly a somewhat futile task to attempt to take stock of a field that is so bent on constant innovation such as digital fabrication. Indeed, it is probable that by the time you are reading this book, new architectural researchers and practitioners will already have updated the origins of digital fabrication in architecture, both conceptual and physical, several times over. Therefore, instead of ending the book with some kind of closing remarks, we conclude with a conversation with three distinct practitioners in computational design and digital fabrication: Mollie Claypool, Dr. Madeline Gannon, and Dr. Behnaz Farahi. By ending the book with further dialogue, we aim to make the case for an evolving narrative of digital fabrication in architecture: one that can constantly shift and update in order to continually orient digital fabrication's rapidly changing coordinates in the field. Such a mutable, contemporary narrative of digital fabrication can also work to resist the kind of singular, top-down histories that were promoted by 20th-century architectural historians, and which continue to dominate architectural education today. These singular, grand narratives tend to privilege the views of one to the exclusion of many. Instead, this book will examine the personal histories of digital fabrication in architecture, who has participated in and been excluded from digital fabrication research, and how these researchers have oriented their work in this emergent subdiscipline. By surveying diverse theoretical starting points, we hope that this volume can be a tool that future digital fabrication researchers, educators, and practitioners can use to critically situate their own work.

And so, in the spirit of initiating such a conceptual reevaluation, let us now revise our original origin point in light of the path we just traversed. When architects home their digital fabrication projects, they determine their conceptual origin points. This homing routine is necessary and productive since it provides them with known historical and theoretical coordinates that will serve as references for all subsequent points they may visit throughout the project's development: without these origins, these architects would quickly get lost.

Homing routine: complete. Now, let's make something.

Notes

1 Anthony Vidler, *Histories of the Immediate Present: Inventing Architectural Modernism* (Cambridge: MIT Press, 2008), 6.
2 Carlo Ginzburg, "Microhistory: Two or Three Things That I Know about It," trans. John Tedeschi and Anne C. Tedeschi, *Critical Inquiry* 20 (Autumn 1993), 10–35.
3 Randy Deutsch, *Superusers: Design Technology Specialists and the Future of Practice* (New York: Routledge, 2019).

Homebound

Theodora Vardouli

Weaving together this book's thrilling collection of hands-on reflection and practice-based speculation on the mélange of practices, tools, materials, techniques, and rhetoric that is digital fabrication in architecture is an enticing metaphor: the "home." Suffice to utter the word and new possibilities rush forward. The editors offer some options: the meanings of "home" can range from technical nomenclature in computer-aided manufacturing (CAM) software to an affective relationship with precedent; its analytical implications can span from ethnographic specificities to historical excavations. Confronted with the "home's" mighty generativity, I here opt to specifically engage the term, together with the evocative act of "homing," as an attitude toward the past. My goal is to activate possibilities for a history of the machine's "home" and its "homing" beyond the tropes of nostalgia and longing for pre-industrial relationships with matter, instruments, and creative agency.[1]

The choice to treat "home" and "homing" as negotiating relationships between the forward-thrusting field of digital fabrication and its elusive ancestors, of course, leaves out other enthralling possibilities. One of them is the specificity of the "home" as an interface metaphor; one akin to the "windows" and "sites," "folders" and "files" that promise to render the alien world of computational objects and functions intuitively navigable and operable. The skeuomorphic approach to the "home" as a technical feature of digital fabrication processes can activate questions around its enactment in material settings and geographies that perturb CAD/CAM software designers' imagined universal isomorphisms; the media sediments that it contains; or the historical decisions that brought the "home" into being in basal CAM projects. I set these questions aside for

DOI: 10.4324/9781003296522-2

now awaiting with eagerness work by fellow scholars who explore the contingent mappings and operational certainties of the "home" metaphors' action-structuring semantics.

Instead, I take on "home" and "homing" as an attitude toward history; as part of the desire to establish a relationship with ever-elusive "origins" or "ancestors" to make sense of new tools and media that enter architectural practices and projections at full force. The editors refer to "homing" as a "longing" and a "method of returning." Such longing is what Antoine Picon diagnoses as "muted nostalgia" in the retrieval of Victorian authors such as John Ruskin and the idolization of the figure of the "craftsman" in contemporary writings about digital fabrication.[2] Picon argues that what is at stake in such invocations is the hope for a renewed relationship with matter. Picon associates digital fabrication with a fundamental shift in the understanding of "materiality," a term that he defines as referring concurrently to matter, to human subjects, and to their relationship.[3] He locates this shift, ushered by scholars like Richard Sennett, Tim Ingold, and the designers who cite them and who they cite, as the desire to establish a *sympathy* with the material world – the very sympathy that the digitalization of matter negates by disembodying matter as information.[4] In nostalgia, Picon sees not only the exigency of disciplinary self-reflection but also an opportunity for "recapturing" and "reinventing" the objects of one's longing.[5]

Mario Carpo writes about digital fabrication, the pinnacle of what he terms "the second digital turn," as the actualization of such reinvention.[6] For example, iterative digital simulations that virtually make and break an object until one reaches an absence of failure (instead of an optimum) are for Carpo a reinvention of pre-industrial craftsmen's trial-and-error practices.[7] Carpo sees the source of Picon's nostalgia, the longing for a sympathy toward a now fully informationalized matter, resolved through current digital fabrication practices. In the "new science of granular prediction" that comes with simulations of material properties and the hyper-precision of digital fabrication machines, he sees the possibility to "liberate" and "almost animate" inorganic matter by establishing protocols for adapting and reacting to unpredictable material events.[8] "Whether we like it or not," Carpo pithily argues, "the future of robotics may be closer to popular quackery than to industrial engineering."[9]

Industrial capitalism figures as a steady backdrop against which to theorize and historicize digital fabrication. In Picon's view, such capitalism might be seen as a historic parallel to the present because of its nostalgia for pre-industrial craft. Alternatively, Carpo may argue that this economic system is a register of the radical shift from prediction to emergence that has taken place upon the technical continuum of calculation and computation. At the core of digital fabrication discourse, and arguments about its relationship with craft traditions, is a fundamental distinction between *fabrication* and *manufacturing* as separate processes. A dictionary search, for instance, delineates manufacturing as a process that is large-scale, repetitive, even unimaginative, and unoriginal.[10] Fabrication, on the other hand, is a form of manufacturing that is about the skillful shaping of materials.[11] Such distinctions carry over to the industry, with a cursory web search differentiating between turning raw materials into products (fabrication) and assembling pre-made parts (manufacturing).[12] Manufacturing lies within the logics of the assembly line, whereas fabrication tends to be portrayed as a more immediate and unpredictable process. We can see such distinctions carrying over to the positioning of CAM and digital fabrication as separate phenomena to be considered in terms of their oppositions rather than alignments – as Carpo does, for instance, when distinguishing the optimizing impulses of industrial engineers from the creative heuristics of digital fabricators.

Industrial Enterprises

Despite these discursive divisions, to which I will return, the titular "machine's" first home was indeed the factory floor, or to be more precise, an expanded site of research, development, and testing under the umbrella of what historians have termed the "industrial-military-academic complex."[13] A detailed account of computer-aided design and numerical control's entanglements with industry-university collaborations and government funding, comes from Daniel Cardoso Llach's examination of the Electronic Systems Laboratory at MIT.[14] Cardoso Llach shows that what is commonly acknowledged as the first numerically controlled machine in 1952 grew out of a collaboration of university researchers with an aircraft company with the aim to automate laborious processes of manual labor for manufacturing airplane parts. Cardoso Llach demonstrates that the need to automate the arduous process for producing the punched paper tapes that controlled the machine led to development of a programming language that abstracted the manufacturing process itself and ultimately to the coupling of computer-aided design with numerical control as intertwined projects.[15] Endeavors to create universal manufacturing machines that could be programmed like digital computers were also taking place outside major technology centers such as MIT. For instance, John Vardalas documents the rise and fall of a Canadian endeavor to create a machine, that went by the name UMAC 5, that could adapt to different machine tools by reprogramming as opposed to hardware and circuitry replacement.[16] Despite the excitement about the potential of a fully programmable digital control system, UMAC 5 encountered a variety of obstacles during its development, sometimes due to computer programmers' unfamiliarity with heavy tools and manufacturing processes that led to various errors and failures. Ultimately, UMAC 5 was perceived as cumbersome and expensive, and Sperry Canada switched to less sophisticated manufacturing systems.

Cases such as UMAC 5 speak to the centrality of distinct historic contexts, with their own funding structures and priorities, in persevering to improve a seemingly failing technology. It speaks to Cardoso Llach's point that computerized numerical control and the project of computer-aided design – the redescription of geometry and design processes in mathematical and computational terms – bootstrapped each other not only to engender a distinct area of research and expertise but also to produce machines that fostered the expectations of efficiency and efficacy attributed to computer-based automation. Efforts in computerized numerical control proliferated across the Atlantic, through distinct networks of academic influence, and were reconfigured as they crossed into different institutional structures and technical infrastructures. For instance, to follow again Cardoso Llach, CAM research in Cambridge, UK engaged the development of rapid prototyping machines for the design of smaller, everyday objects.[17]

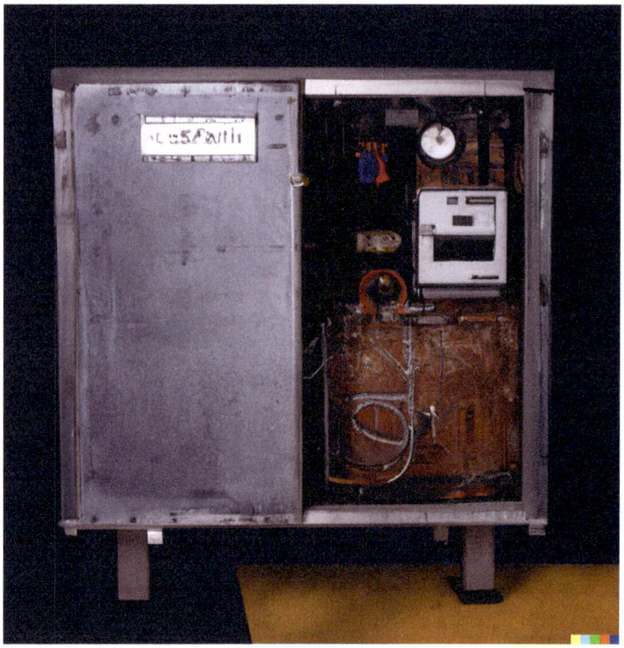

Figure 2.1 Image of a fictional machine generated with OpenAI DALL-E based on the input of an archival photograph of the historic numerical control machine UMAC 5 from Ingenium, Canada's Museums of Science and Innovation (original photograph number 1985.0359.001).

Although these contexts seem far removed from contemporary digital fabrication settings and technics, we may pause to consider for a moment that the university research laboratory persists as a privileged site of making with computationally controlled machines and processes. Many celebrated innovators in architectural digital fabrication, such as the ICD Stuttgart or the Mediated Matter Group and the Self-Assembly Lab at MIT to name a few, operate as university research laboratories rehearsing modes of interdisciplinarity and funding mechanisms that echo from basal computerized numerical control projects.[18] Because of their size, cost, and requisite skill for operating, digital fabrication machinery is still mostly bound to the sites that digital computers used to occupy before appearing on everyone's desktops and backpacks: research laboratories and offices (in our case architectural studios) that have the necessary space, funds, and personnel to maintain and run them. The use of robotic arms and laser cutters in Architecture School fabrication laboratories brings memories of times when computers occupied entire rooms and users had to adhere to strict protocols mandating time and order of use (Figure 2.1).

Personal Fictions

Countering digital fabrication's confinement to the Lab, and with deliberate echoes of digital electronic computers' historic move from the lab to the home, are efforts to make digital fabrication technologies truly *personal*.[19] This emergent counterculture remains a somewhat underdiscussed topic in architecture, in stark contrast with ample discourse and optimism about digital design's democratizing and emancipatory potential that I have discussed elsewhere.[20] This delay is likely an asset, as architects and designers can afford a critical distance from "fab labs," "hacker spaces," and "maker communities" in order to reflect on some of their thornier issues; not to reject them as visions but rather to venture cautiously toward them.

A foremost advocate for democratizing digital fabrication has been MIT Media Arts and Sciences professor Neil Gershenfeld. His almost 20-year-old book *Fab: The Coming Revolution on Your Desktop* described a world akin to what futurologist Alvin Toffler prophesied in his *Third Wave* – a world of productive consumption or "prosumption."[21] Digital fabrication machines, now desktop-sized and easier to use through the support of resource and expertise-sharing communities such as the one he set up in his class *How to Make (Almost) Everything* at the MIT Media Lab, would allow for *personal* fabrication. Personal fabrication did not only mean making objects for one's own use, such as 3D-printing a coffee cup, but also channeling imagination and creativity; a trope nostalgic of the early years of computing and the internet. Gershenfeld also sermonized bolder visions, a machine that could make other machines, and ultimately matter itself, as programmable material capable of computation – oddly a more familiar vision in architectural digital fabrication than the rest of the book's propositions.[22]

Gershenfeld's book had a wide reach. Testament to its readership and audiences was a review by renowned post-phenomenologist philosopher Don Ihde for the major science and technology journal *Nature*.[23] Ihde's review scathed the ushering of yet another "revolution" and the book's "technofantastical" tone but found one of its aspects to be "infectious" and "possibly of revolutionary importance."[24] This was the way that Gershenfeld had "exported" and "prepared for use by children," in Ihde's terms, a packaged piece of digital fabrication infrastructure known as a "Fab Lab." In 2006, Gershenfeld was describing fab labs as small-room-sized places consisting of fabrication equipment that required about $30,000 to set up but was projecting a drop in the cost to $10,000 or even $1,000 once machines would become smaller and more affordable.[25] In 2016, the global network of Fab Labs was counting about 1,000 facilities worldwide.[26]

Ihde saw the promise of fab labs as one of technology democratization and of self-learned skills through eight-year-old children playing with the machines.[27] The democratization promise has dominated many accounts of fab labs, and the so-called "maker movement" that they have supported and instigated. The Fab Lab network grows together with a distributed network of learning resources, the "Fab Academy," that offers the required training and skills.[28] Like other democratizing narratives associated with digital technologies, global digital making comes with its own contentious histories and social visions. Sarah R. Davies has traced the fascination with "making" and personal fabrication to DYI and active leisure cultures, to ultimately argue that argue that enthusiasm about the maker movement is not just about new technologies becoming available, but instead carries with it a specific vision of citizenship.[29] This vision, she argues, revolves around self-governance, enterprising, and individual responsibility.[30] The figure of the empowered and aware individual – a figure whose construction is entangled with techno-utopianist tropes advanced within American counterculture – overhauls other modes of political agency that revolve around collective organizing and action.[31] This proposition becomes especially knotty if we pause to consider Fab Labs as an American "export" that is disseminated across the globe and adopted by various communities.[32] The dissemination of fab labs confronts us with the oxymoron of a *universal* project of *personal* fabrication and ultimately the specter of technocolonialism.

Colonial Frictions

Cardoso Llach and Andres Burbano warn us that "adoption" approaches to digital technologies risk framing communities in terms of their fitness to serve predetermined technological projects.[33] Morgan Ames' work on the One Laptop Per Child project – MIT Media Lab founder Nicholas Negroponte's endeavor to disseminate laptops across children in the Global South – shows clearly that visions of universal digital literacy as a socially transformative project are part and parcel of colonial attitudes.[34] Syed Mustafa Ali even more radically suggests that computing "is colonial through and through," because of its reliance on the project of Western modernity and its ways of knowing and being based on systems of categorization, classification, and taxonomization and their manifestations in practices, artifacts, and technologies.[35]

As a response, which is also relevant for our (re)thinking of personal fabrication, Ali proposes "decolonial computing" as a critical framework and an ethic. Decolonial computing radically includes the subjects, epistemologies, and ontologies situated at the peripheries of the world system and aiming to decenter of Western-centric universals.[36] Another useful framework is what Kavita Philip, Lily Irani, and Paul Dourish refer to as "Postcolonial Computing" – a set of tactics for flipping the script, as they put it, on hegemonic phenomena bundled under that very same term.[37] Their subversion offers a list of useful principles, amongst them the rejection of schemas of technology "export" and "adoption" and the recognition of technology and design as irreducibly situated activities. Especially relevant amongst their observations is the dissolution of a fantasy of binary opposition between Western science and reason and non-Western cultural practices, which were always untenable but are most obviously so in the context of the "international traffic in design discourses, technologies, and people."[38]

Where to Next

As we think about decolonial and postcolonial futures of digital fabrication, as we seek new homes for the machine in critical juxtaposition with its past conceptual residences and technosocial sites, we need to think about *access*. Access is not only about putting something within one's reach. It is a complex notion that has steadily figured as a bone of contention in social and political projects that positioned computing and digital technologies as devices for individual and collective empowerment such as the Free/Libre

Open Source community. For instance, Free Software herald Richard Stallman has often criticized Eric Raymond's definition of "open source" for foregrounding the openness and shareability of code, without specifying who and under what conditions said code is made available to read and modify – i.e., without considering the question of access.[39]

Vernelle Noel, Yana Boeva, and Hayri Dortdivanlioglu's recent work on the knotty question of access in digital fabrication offers inspiring directions.[40] Their work suggests that digital fabrication is already one step closer to a more equitable future because it elevates and dignifies making in a discipline and profession whose Western origin stories and Albertian distinctions between the intellectual task of designing architecture and the manual task of constructing a building have deliberately excluded it.[41] Through participant observation and auto-ethnography, they show that digital fabrication training in educational and institutional settings privileges gendered and colonial tropes of technical expertise, disembodiment, accuracy, and precision. To it, they juxtapose what they refer to as *enskilment* through craft, that is, a situated and materially embedded process of building skills in a specific craft practice – in this case wire bending from the Trinidad carnival which Noel researches and makes with – by being attuned to the social, corporeal, and material entanglements of the work. In Noel's work, computation becomes a method of encoding a culturally rich craft practice that is dying with its masters to reenact it in new settings.[42] Instead of being a requisite from the outset, digital fabrication machines are introduced later in the process as yet another tool to mediate relationships between the body, materials, and others.[43]

Computational making offers an alternative category that includes but is not limited to digital fabrication, by decentering the machine and considering relationships between mathematical abstraction and materially embedded and embodied making.[44] Shelby Doyle's work at the Iowa State University Computation and Construction Lab, which also champions equity, tactically uses "computation" instead of "digital" to point to technocultural regimes rather than robotic arms, 3D printers, and CNC mills.[45] Felecia Davis' work in computational textiles and Dina El-Zanfaly's research on improvisation in making with machines adopt similar sensibilities. Together with emerging allied efforts, these works draw from pasts, experiences, and traditions that have been systematically excluded from the historical development of digital fabrication machines and processes as high-technology instruments for automating manual processes.

"Homing," in these works, evades longing and nostalgia and becomes a critical project and a program of action. We turn the eye toward the past because of a lingering suspicion that, maybe, the conditions of digital fabrication's elusive and precarious "origin" haunts current practices and sway its future orientations. Walter Benjamin's take on Klee's "Angelus Novus" as an allegory of progress comes to mind: eyes toward the past and back turned against a future toward which he is irresistibly propelled by the "storm of progress." Meanwhile, "the pile of debris before him grows skyward."[46] To "home" the machine means to negate the techno-optimism of progress narratives and contend with the debris. It also means to seek a new home for the machine – a home where the machine is not the center.

Notes

1 For accounts of digital fabrication and nostalgia, which I also discussed later in the essay, see Picon, Antoine, 2019. Digital Fabrication, Materiality and Nostalgia. In Retsin, Gilles, Jimenez, Manuel, Claypool, Mollie and Soler, Vicente. *Robotic Building: Architecture in the Age of Automation*, München: DETAIL. https://doi-org.proxy3.library.mcgill.ca/10.11129/9783955534257; Carpo, Mario. 2017. *The Second Digital Turn: Design Beyond Intelligence.* Writing Architecture. Cambridge, MA: MIT Press.

2 Picon, Antoine, 2019. Digital Fabrication, Materiality and Nostalgia. In Retsin, Gilles, Jimenez, Manuel, Claypool, Mollie and Soler, Vicente. *Robotic Building: Architecture in the Age of Automation*, München: DETAIL. https://doi-org.proxy3.library.mcgill.ca/10.11129/9783955534257, 112.

3 Ibid., 113.
This relational definition of materiality also appears in Katherine Hayles' distinction with what she refers to as "physicality." Materiality, she argues is a dynamic process that shifts together with the focus of attention. (Hayles, Katherine N. 2012. *How We Think: Digital Media and Contemporary Technogenesis.* Chicago: University of Chicago Press).

4 Ibid.

5 Ibid., 114.

6 Carpo, Mario. 2017. *The Second Digital Turn: Design Beyond Intelligence.* Writing Architecture. Cambridge, MA: MIT Press.

7 Ibid., 46.

8 Ibid., 50.

9 Ibid., 55.

10 "manufacture, v.." OED Online. March 2023. Oxford University Press. https://www-oed-com.proxy3.library.mcgill.ca/view/Entry/113770 (accessed April 18, 2023).

11 "fabricate, v.." OED Online. March 2023. Oxford University Press. https://www-oed-com.proxy3.library.mcgill.ca/view/Entry/67397?rskey=AN1WcW&result=2&isAdvanced=false (accessed April 18, 2023).

12 See for example, What Is the Difference Between Manufacturing and Fabrication. https://www.optiproerp.com/blog/what-is-the-difference-between-manufacturing-and-fabrication/ (accessed April 18, 2023) and Manufacturing vs. Fabrication: What is the Difference? https://www.pacific-research.com/manufacturing-vs-fabrication-what-is-the-difference-prl/#:~:text=Fabrication%20is%20about%20the%20creation,suited%20to%20a%20specific%20project (accessed April 18, 2023).

13 See, for example, Leslie, Stuart W. 1993. *The Cold War and American Science: The Military-Industrial-Academic Complex at MIT and Stanford.* New York: Columbia University Press.

14 Cardoso Llach, Daniel. 2015. "Software Comes to Matter: Toward a Material History of Computational Design." *Design Issues* 31 (3): 41–54. https://doi.org/10.1162/DESI_a_00337.

15 Ibid.

16 Vardalas, *The Computer Revolution in Canada.*

17 Cardoso Llach, D., & Forrest, R. 2017. "Of Algorithms, Buildings and Fighter Jets: A Conversation with Robin Forrest." *Architectural Research Quarterly* 21 (1): 53–64. https://doi.org/10.1017/S1359135517000173.

18 On architects' entanglements with science and technology in the research university see, for example, Dutta, A. 2013. *A Second Modernism: MIT, Architecture, and the "Techno-Social" Moment.* Cambridge, MA: The MIT Press and Sachs, Avigail. 2018. *Environmental Design: Architecture, Politics, and Science in Postwar America.* Midcentury: Architecture, Landscape, Urbanism, and Design. Charlottesville: University of Virginia Press.

19 Chandler, David. 2016. 3 Questions: Neil Gershenfeld and the spread of Fab Labs. URL: https://news.mit.edu/2016/3-questions-neil-gershenfeld-fab-labs-0104

20 Theodora Vardouli and Olga Touloumi eds. Computer Architectures: Constructing the Common Ground. New York: Routledge. 2020.

21 Gershenfeld, Neil. 2007. *Fab: The Coming Revolution on Your Desktop--From Personal Computers to Personal Fabrication.* New York: Basic Books; Toffler, Alvin. 1980. *The Third Wave.* New York: Morrow.

22 On programable matter, see Tibbits, Skylar. 2021. *Things Fall Together: A Guide to the New Materials Revolution.* Princeton: Princeton University Press.

23 Ihde, Don. 2005. "Making Progress?" *Nature* 435: 23. https://www.nature.com/articles/435023a.

24 Ibid.

25 2006. "Fab Lab." *MIT Spectrum.* https://spectrum.mit.edu/spring-2006/fab-lab/.

26 Chandler, 3 Questions.

27 Ihde, Making Progress?

28 Gershenfeld Neil and Jim Euchner. 2015. "Atoms and Bits: Rethinking Manufacturing." *Research-Technology Management* 58 (5): 16–23. https://doi.org/10.5437/08956308X5805003.

29 Davies, Sarah R. 2017. *Hacker Spaces: Making the Maker Movement.* Polity, 151.

30 Ibid.

31 Ibid., 147.

32 In their otherwise optimistic account, Laure Morel and Serge Le Roux have pointed out that despite the world-wide spread of Fab Labs, the United States has played and is still playing a major role in their dissemination.
Morel Laure, and Serge Le Roux. 2016. Fab Labs: Innovative User. ISTE Ltd.
33 Cardoso Llach, Daniel and Andrés Burbano. 2020. "Other Computations: Digital Technologies for Architecture from the Global South". *Dearq* (27): 6–19. https://doi.org/10.18389/dearq27.2020.01.
34 Ames, Morgan. 2019. *The Charisma Machine: The Life, Death, and Legacy of One Laptop per Child.* Cambridge, MA: The MIT Press.
35 Ali, Syed Mustafa. 2016. "A Brief Introduction to Decolonial Computing." *XRDS: Crossroads, The ACM Magazine for Students* 22 (45): 16–21. https://doi.org/10.1145/2930886.
36 Ibid., 21.
37 Philip, Kavita, Irani, Lilly, and Paul Dourish. 2010. "Postcolonial Computing: A Tactical Survey." *Science, Technology, & Human Values* 37 (1): 3–29.
38 Ibid., 23.
39 Vardouli, Theodora, and Leah Buechley. 2014. "Open Source Architecture: An Exploration of Source Code and Access in Architectural Design." *Leonardo* 47(1): 51–55. http://www.jstor.org/stable/43834115.
40 Noel, Vernelle AA, Boeva, Yana, and Hayri Dortdivanlioglu. 2021. "The Question of Access: Toward an Equitable Future of Computational Design." *IJAC* 19 (4). https://doi.org/10.1177/14780771211025.
41 Ibid.
42 See, for example, Noel, Vernelle AA. 2022. "Computational Regionalism: De-familiarization of Tectonics in the Wire-bending Craft." *IJAC* 20 (2): 277–296.
43 Noel et al. The Question of Access.
44 On the conceptual and technical aspects of computational making see, Knight Terry and Theodora Vardouli. 2015. Special Issue: Computational Making. *Design Studies*. 41 (Part A).
45 See Shelby Doyle's chapter in this book.
46 Benjamin, Walter. 1942. "On the Concept of History,"

The Technological

As we now watch robotic quadrupeds prowl construction sites and 3D-printed goo congeal into houses, something has become clear for those in the world of digital fabrication research: the medium has become the message. In other words, architectural technology has become so awe-inspiring that, much like Marshall McLuhan's canonical phrase suggests, content is less important than technology itself. This technological takeover can be seen in mainstream discussions on robotics, generative design, and artificial intelligence, in which engineering feats tend to precede their real-world applications (if the latter is mentioned at all). In academia, for instance, conferences are generally categorized by tools rather than objectives, reinforcing the notion that the tool—and not its multiplicity of potential values—is the protagonist in the narrative of digital fabrication research and practice.

This phenomenon is, of course, not new. As many scholars have acknowledged, revolutions are often tied to technology. Even McLuhan's axiom emerged out of the so-called mass media revolution. Is it possible then to decenter technology in digital fabrication discourses? Must it always be regarded as a heroic device? An alternative perspective on this centering of technology might come from architect Cedric Price, who famously quipped, "Technology is the answer, but what was the question?" With this provocation, Price inspired people to critique the march of technological progress and question its ostensibly revolutionary message. While Price acknowledged that that technology is indeed the message, he also encouraged skepticism of this heroic position. The question is what matters.

DOI: 10.4324/9781003296522-3

The discussions in this section ask questions to which technology may be the answer. They explore the cultural aspect of today's digital fabrication gizmos and gadgets. Rather than placing the technology at the center, however, the subject is tackled obliquely from the points of view of history, academic critique, and self-reflection. Here, the goal is to challenge the premise that the architectural sub-discipline of digital fabrication is solely about being high-tech. Digital fabricators find themselves in this field for a variety of reasons, ranging from a simple interest in being cutting-edge to a desire to "design the future." Practices are personal and fraught with different dynamics between people and machines. Asking questions, as Price suggested, is a form of recalibrating these relationships.

The technological is the obvious and most common lens through which to view digital fabrication in architecture. For our purposes, it is simply the starting point. The technological perspective will launch other dialogues on tools, materials, history, people, culture, and society that cannot be easily resolved yet shed light on the realities of working in this field. We may find that technology is not *the* message but simply one of many messages.

Lost and Found

In Search of a Process for Creating Productive Problems

Michael Szivos

Introduction

Unlike case studies typically found in the emerging discourse of digital fabrication in architecture, this case study is not on a single, technical project. Instead, this case study is on process, specifically, a process of conceptual exploration that began with the founding of our design and fabrication studio, SOFTlab, and continues with its work today. Our goal in retracing this process is not to mark its achievements, but rather to expose the errors and naïve assumptions we made through its evolution and, further, to reflect on how such miscalibrations became "productive problems" that opened new possibilities for our work. Through such reflection, we aim to distill a new process for purposefully creating productive problems that we can draw on for future invention.

Naive Curiosity: Production Problems as a Speculative Framework, 2005–2007

I founded SOFTlab in 2005 shortly after graduating from Columbia University's Graduate School of Architecture, Planning and Preservation (GSAPP) with a Master of Science degree in Advanced Architectural Design. The earliest explorations of the studio were driven by an enthusiasm toward solving problems of mass customization by automating production workflows with custom software and new digital fabrication technologies. For

DOI: 10.4324/9781003296522-4

instance, our very first project, which I began as an independent study with other graduate students at GSAPP, stemmed from a seemingly simple question: how can we use computation to automate a fabrication workflow to build a complex surface? At the time, a "complex surface" for us, and most other nascent digital makers, meant a doubly curved surface that, when rationalized, resulted in a collection of planar panels that were all unique from one another. This relatively new understanding of complexity reflected a collective, disciplinary shift: with the advent of digital tools, architects were no longer merely interested in designing complex forms, but rather in designing the complex systems that could rationalize such forms for construction. The computational workflow we aimed to develop was thus not only for designers to rationalize complex surfaces into planar panels but also to create toolpaths for CNC fabrication of these panels, as well as panel organization and labeling systems. Similar workflows are now conveniently packaged into readily available CAD/CAM software, such as Autodesk Fusion 360 or SolidWorks, but we had no such tools at the time.

In this first project, and indeed in most of our subsequent projects, we made a series of initial, basic decisions; in doing so, we unknowingly invited a number of multilayered questions to ask and additional decisions to make. Our first decision was to divide the workflow development of this project into two tracks: the coding required to automate the rationalization process and the design of the construction system and details. We realized early on that one benefit of automation was that we could easily plug-in a range of surface geometries, therefore the design of the overall geometric form was thus not as immediately important to us as its construction details. We developed physical prototypes of the construction details (Figure 3.1) while we developed the automation code; our thinking was that we could simply plug in the final iteration of the details once the code was complete (Figure 3.2).

We coded the automation script in Autodesk Maya using MEL (Maya Embedded Language). We made particularly extensive use of the "history" and "nodes" inherent to

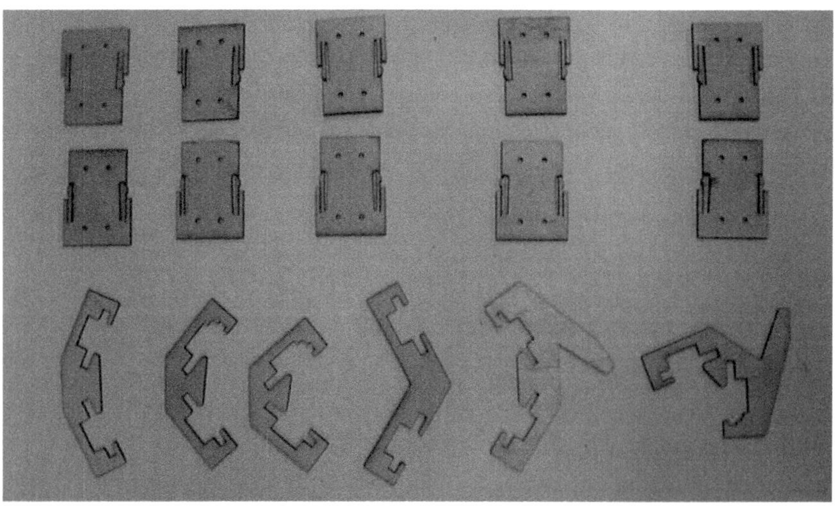

Figure 3.1 Basswood laser-cut joint prototypes to test for tolerances and fit.

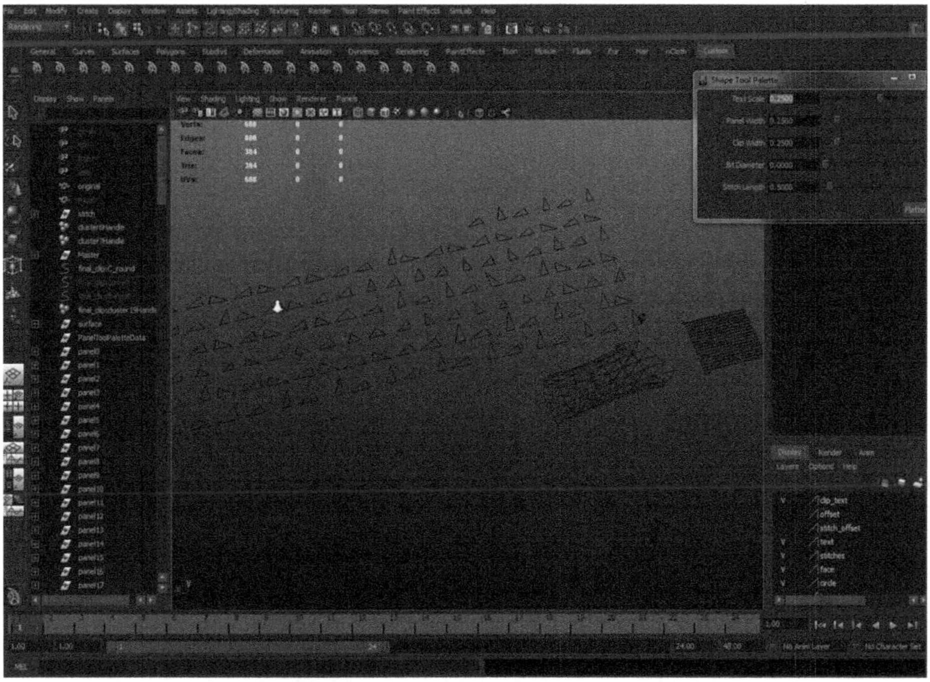

Figure 3.2 Screen capture of executed custom script in Maya to rationalize the geometry and automate the production of labeling and tool paths.

Maya's animation and modeling platform.[1] At GSAPP, we had access to CNC equipment that could quickly and precisely cut sheet material, we therefore decided to use our Maya scripts to output tooling and/or line work for two-dimensional CNC cutting. This decision ultimately led to the goal of constructing the complex surface strictly out of panels rather than a stick frame system, which would require separate structural members and cladding; in this project, the structure and the cladding would become one in the same.

Our leap from computational workflow to physical construction produced several unanticipated problems, many of which led to discoveries and lines of inquiry that we continue today. For example, when we set out to automate the production files for a "self-supporting surface" we didn't really consider what "self-supporting" meant. We simply assumed that we would forgo any kind of structural framing by constructing a surface out of CNC-cut plywood panels and custom laser-cut steel joints (Figure 3.3) that precisely registered the angles between panels. Since we did not have framing, we assumed that our self-supporting surface assembly needed to be made of strong sheet materials: the panels were made from plywood and the joints were made from either steel or aluminum.

These naïve design assumptions had a number of material consequences. First, the plywood panels were too heavy for the metal joints, causing them to torque and deform. The plywood, in turn, was too soft to resist the torquing metal joints and was quickly pulled apart. Second, since we aimed to literally translate the folded digital form into an assemblage of plywood panels, we did not fully consider that the unsupported

Figure 3.3 Prototype detail made of plywood and steel joints.

cantilevers that were possible in the gravity-less digital environment would not be possible in the world of physics. Our initial prototypes quickly started to fail: the weight of the plywood panels proved to be too great and the steel joints tore through the soft plywood slots they were inserted into.

In this first prototype, we were more concerned with writing code to automate the rationalization, labeling, and tool path planning for complex surface panelization rather than the ways in which the design and construction of complex surfaces can impact their structural integrity. Looking back I think, we were distracted by the automation process and the novelty of working completely in digital space. We just assumed that the panels and joints would be rigid enough to support any complex surface. What we discovered was that we had a lot to learn about how to account for material and structural performance in the design of digital fabrication workflows.

Rethinking: Material Feedback and Prototypes as Projects, 2008–2009

One of the main issues with the fabrication of our initial prototype was the weight of the material. As a result, we started to experiment with lighter materials like cardboard. However, we knew that cardboard would not have the strength of the previous materials we tested. We therefore also begin to think about how the overall surface geometry could be inherently structural.

For instance, we learned from our first attempt that the surface should not have any creases or folds where the structural loads were not distributed uniformly throughout the surface. However, we didn't know how to design a surface that automatically distributed loads evenly. We started looking at tensile surface systems and eventually began to develop surfaces using Maya's built-in cloth and physics simulation tools. We ultimately found "minimal surfaces" and the computational form-finding techniques capable of creating them.

At the time we didn't think to Google mathematical concepts like "tensile surfaces" or "minimal surfaces." We didn't even really know we were creating minimal surfaces until years later! I think this is worth mentioning to demonstrate that we didn't work with these forms simply because we *liked* them. Rather, we worked with these forms because they emerged as a pragmatic and necessary solution to our goal of building self-supporting surfaces. In other words, what we had initially thought of as a problem of connections and material properties revealed itself to be a problem more of surface geometry. We realized that we needed to design the complex surface just as we needed to design the automation process that would fabricate it.

The first test of our minimal surface fabrication system was a commissioned project rather than a prototype. We were invited to participate in a group exhibition titled "system:system" in what was formerly Saint Cecilia's Convent in Greenpoint, Brooklyn. The exhibition was curated by Christina Vassallo and Adam Henry and included 25 site-specific installations by artists and designers that were centered on the general theme of "rule-based work." The old convent had previously been used for several group art shows and lent itself to installations because it was a building composed of small rooms.

We were assigned a room that had two entrances and two windows. Our installation was a surface designed to connect these separate thresholds and apertures (Figure 3.4). *pAlice*, as we would later call it,[2] was our first built installation (Figure 3.5).

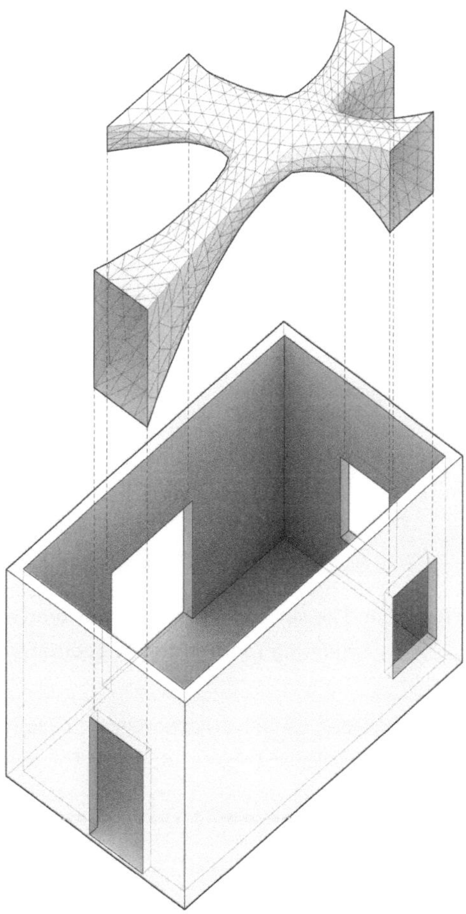

Figure 3.4 Diagram showing how *pAlice* connects the thresholds of the install site.

Figure 3.5 ***pAlice*** **was made of laser-cut chipboard panels and custom joints.**

Figure 3.6 The surface was clad in mirrored mylar panels to reflect the image of the existing room onto the geometry of the installation.

pAlice provided us with an opportunity to use the characteristics of minimal surfaces as drivers of particular design decisions. For example, minimal surfaces are structurally stable, but they also have interesting topological conditions that confuse the registration of inside and outside. We thought this unique geometric trait would be interesting to explore in a room that had so many ways to look into, or through, it.

We imagined the installation as an inversion of the room. All existing apertures but one were connected by the surface so when one looked into the room, they were actually looking into the interior of the installation. Conversely, when visitors entered the room through the one existing opening that had not been absorbed by the structure, they were confronted with the exterior of the installation. We clad the installation's exterior with mirrored mylar (Figure 3.6) in order to reflect the surrounding room and, thus, further confuse the existing and new geometries.

In pAlice, we were able to build a structurally stable complex surface completely out of cardboard. We could thus infer that the panel material's self-weight and the structure's global form were more important design considerations than the panel material's strength. Another unexpected finding from this installation was that a minimal surface can naturally "find" its form as it is being assembled. Therefore, the need for custom joints that precisely register angles between panels was not as critical as we originally considered it to be (Figure 3.7). With this learning in mind, we began to search for more universal, off-the-shelf details that could attach adjacent panels in simpler ways. In pAlice, our fabrication process once again revealed inherent characteristics of the surface we were fabricating, which, in turn, enabled us to further edit and refine our design and fabrication system.

Flexibility and Resolution: Construction Material as a Medium, 2010–2013

The next major iteration of our research into automated fabrication workflows involved two major changes: first, we eliminated the custom joints between panels and, second, we moved away from custom MEL coding to using Grasshopper, a plug-in for Rhino that had just recently been introduced. As discussed earlier, pAlice helped us to understand that minimal surface geometries could naturally assume their forms and, as a result, that we didn't need custom joints to force each surface panel into place. Once we identified that rigid joints were no longer necessary, we were able to test lighter and more flexible materials like polyester and mylar film (Figure 3.8). These sheet materials lent themselves to simpler connections between panels and presented us with opportunities to experiment with colors and optics (Figure 3.9).

We found compelling ways to map color onto each panel but generally felt that we were limited to essentially treating each individual panel as a large pixel. We immediately recognized this limit as a potentially productive problem. We therefore set out to make another minimal surface installation that spatialized a true color gradient; to do so, we turned our attention to using off-the-shelf printer paper as our building material.

The first thing we realized about building with paper was that we needed to rethink our preconceptions about making structural connections. At first, we attempted to add holes and grommets or rivets to the paper panels in order to connect them with one another, but one small hole in the paper would quickly lead to a full tear. After many failed tests, and lots of frustration, one of our team members jokingly folded two sheets of paper and clamped them together with a binder clip that was sitting on their desk. Though the binder clip solution was initially offered in jest, we soon realized that it was precisely the solution we were looking for—binder clips are used to hold stacks of paper together after all! Part of what made the binder clip so effective was that it used friction rather than a mechanical connection, e.g., a hole and a rivet.

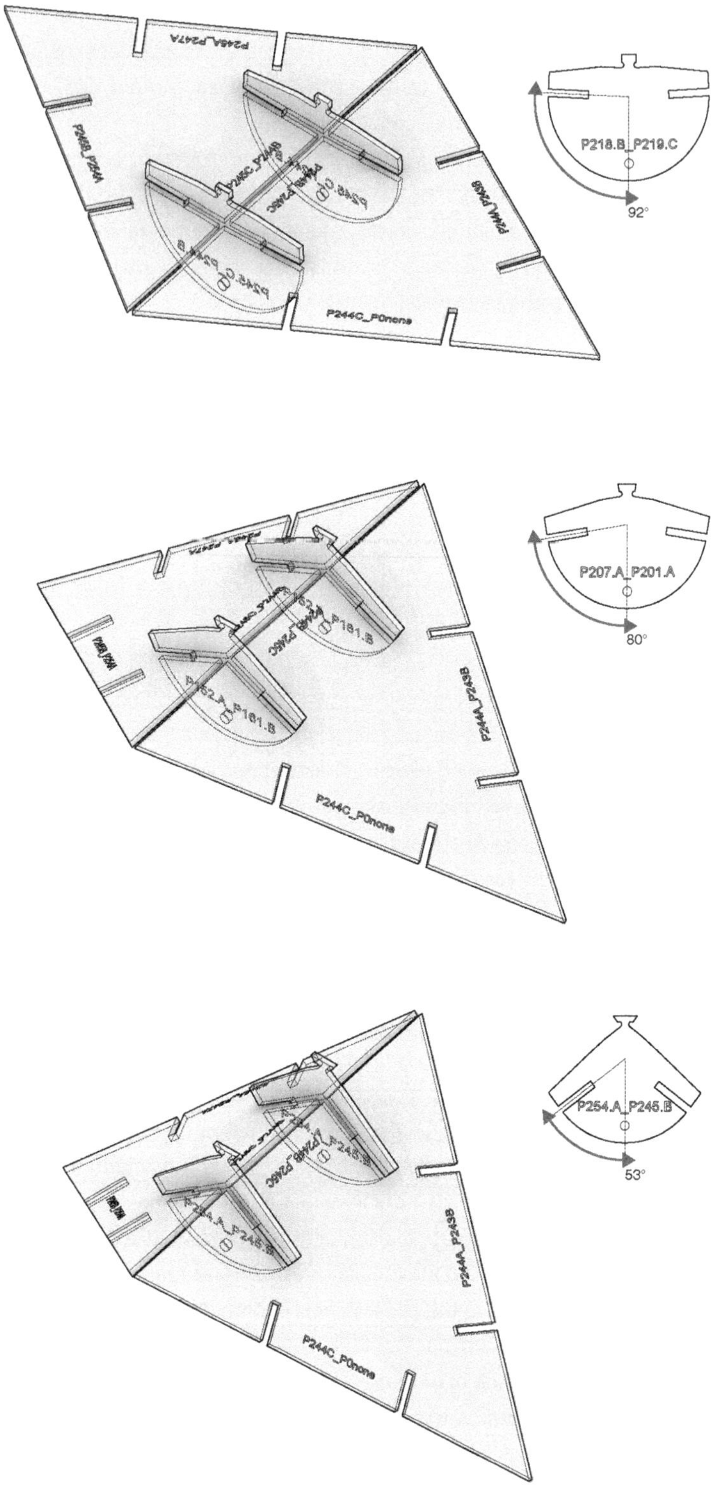

Figure 3.7 Drawings showing how the articulated joints registered the angle between panels.

Figure 3.8 *(n)arcissusi* installed in the stairwell at the Frankfurter Kunstverein in Frankfurt, Germany for the group show "Abstrakt" in 2010.

Figure 3.9 Xtra Moenia installed on Mott Street in Manhattan as the North Gate for the San Gennaro festival in 2010.

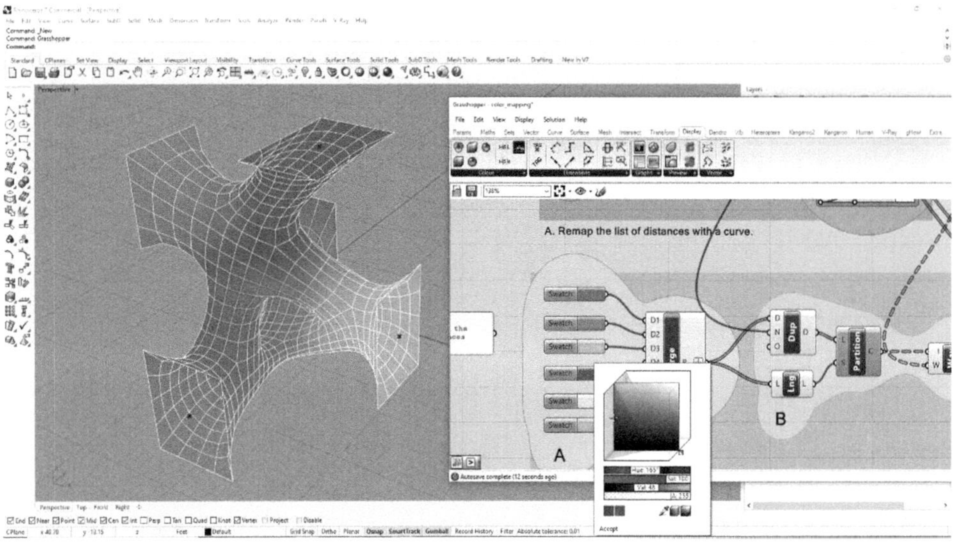

Figure 3.10 Screen capture of Rhino and Grasshopper definition showing the mapping of color onto a complex surface.

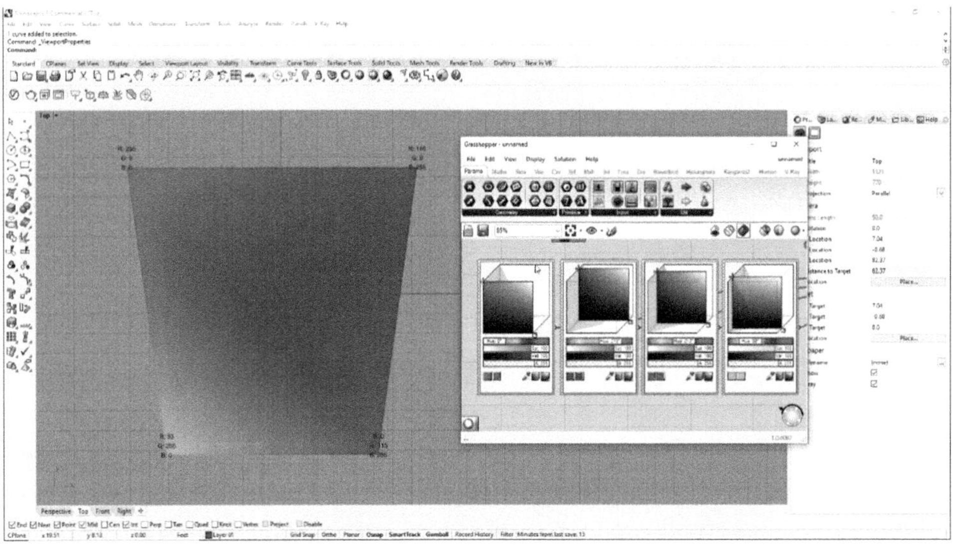

Figure 3.11 Screen capture of Rhino and Grasshopper definition showing how a gradient of color is interpolated between the vertices of a single mesh face.

Now that we had solved the connection problem, we needed to develop a way to map and transfer the color gradient to the surface. We used an attractor point strategy wherein several attractor points were placed in the digital environment and each one was associated with a different color. Each vertex in the surface was mapped with a mix of colors based on its average distance from these points. This gave us a way of mapping the surface three-dimensionally (Figure 3.10). We then transferred the gradient to each panel by interpolating the four colors mapped to each vertex of a given panel (Figure 3.11).

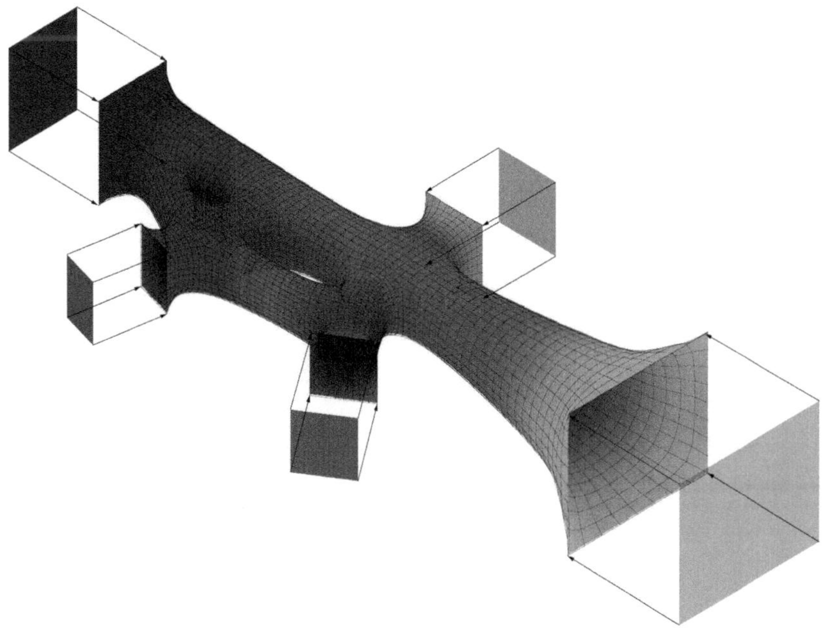

Figure 3.12 Diagram showing how the apertures of *CHROMAtex* each correspond to a color that is projected into the interior of the installation and mixed with the other colors.

We eventually decided to create a large-scale installation in which we would strive to conceptually use color as a building material. By mapping an image of color onto a complex, digitally designed surface, we aimed to blur the lines between constructed spaces and the typical two-dimensional, digital representation of those spaces, i.e., rendered images. We wanted to both critique the then-popular notion of renderings as ideal means of geometric speculation and introduce a new productive problem into our workflow. Just as in pAlice, our conceptual design ideas were informed by the technical production problems we chose to pursue.

We developed our color gradient workflow through a site-specific installation called *CHROMAtex*, which was exhibited at the bridge gallery in the Lower East Side of Manhattan in 2010. The gallery had a large single-pane window facing the street. We decided to use this window as a means to ground the project in the space and as an excuse to revisit our interests in creating ambiguity between the inside and outside of a complex surface, which we first explored in pAlice.

The installation was conceptualized as five apertures, each with a corresponding color (Figure 3.12). The colors of the apertures were mixed to create a spatialized gradient mapped onto the surface connecting the apertures. That surface was then subdivided into panels for construction. Each surface panel in the installation was mapped as an interpolated gradient that aligned with its corresponding cut files (Figure 3.13). The largest aperture of the installation was attached to the window in the front of the gallery so pedestrians on the street could only see the interior of the installation (Figure 3.14).

The laser-cut and print files of each surface panel were mirrored: each piece of paper was first fed through the printer, where it received its unique color gradient, and

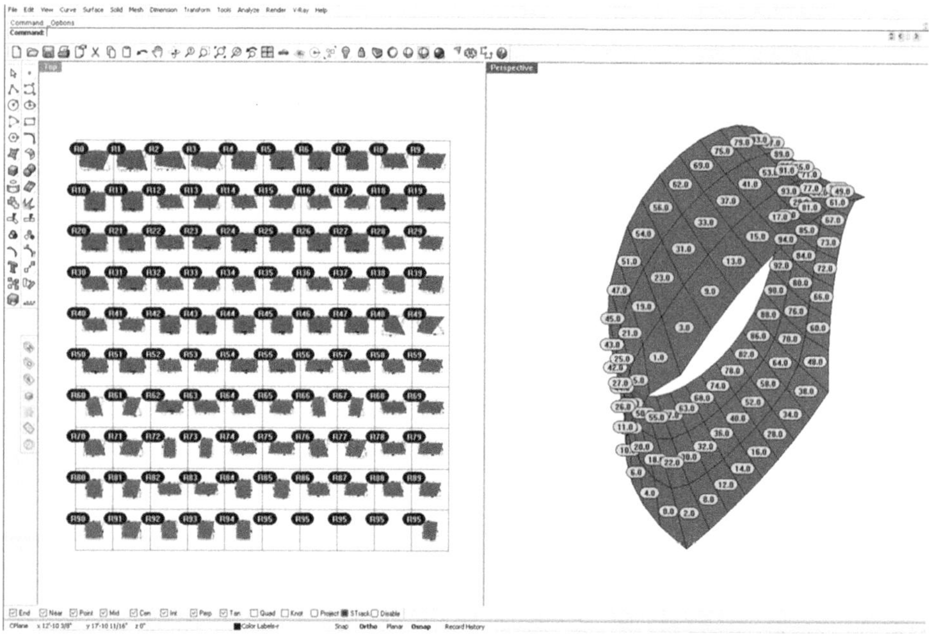

Figure 3.13 Screen capture of Rhino showing a section of *CHROMAtex* and the corresponding flattened panels with labels, tool pathing, and color map.

Figure 3.14 The main aperture of *CHROMAtex* attached to the streetfront gallery window on Orchard Street.

then flipped onto its face to be laser cut (Figure 3.15). Each panel was connected to adjacent panels with four binder clips (Figure 3.16). There were over 17,000 binder clips used in the installation in total.

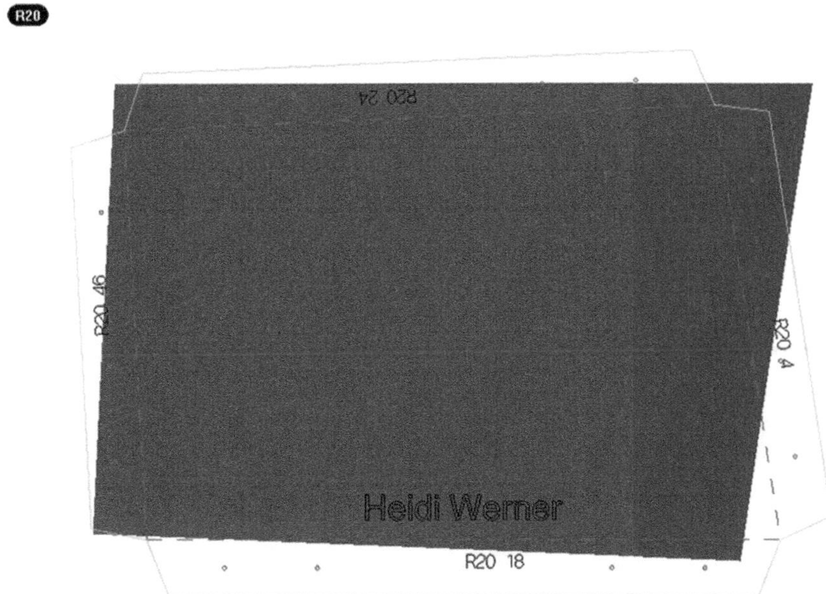

Figure 3.15 File for a single panel showing the color map to be printed and the tool paths for laser cutting.

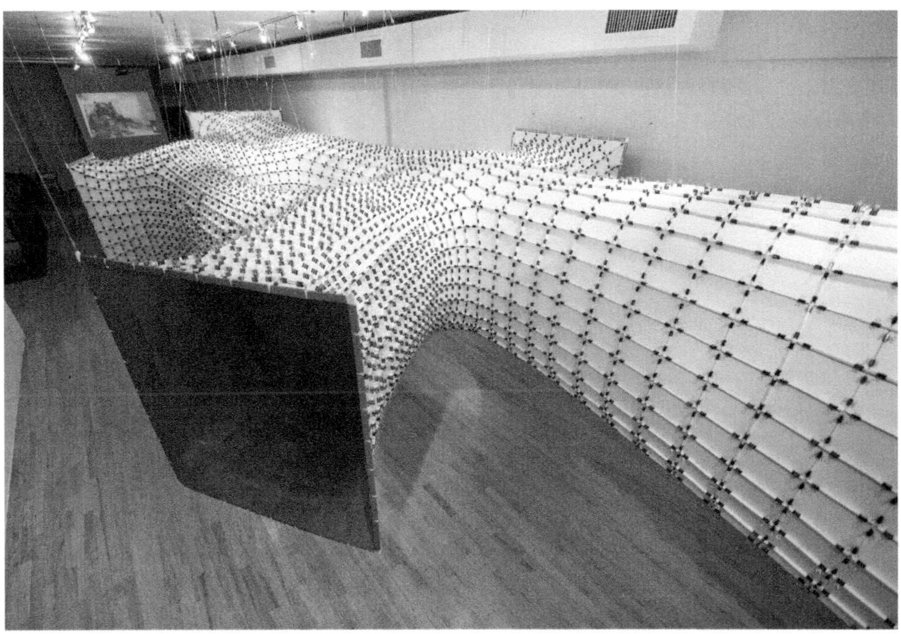

Figure 3.16 *CHROMAtex* was made of laser-cut photo glossy paper connected with binder clips.

The interior of the installation is a 30-foot-long printed image at 300 dots per inch. The three-dimensional mapping of the five colors on the interior produces a spatial blending of surface geometry and color that is seen differently through each aperture (Figure 3.17). The process of mapping the color gradient based on vertices and the different orientations of their respective normals inspired us to speculate on how we might begin to make the panels three-dimensional or composites of multiple materials.

We also recognized that the binder clip detail had an impact on how people perceived the installation. The installation was very much the outcome of a digital, robotic process that can be esoteric to most people, but the everyday quality of the binder clip connections made the piece much more accessible. Many people actually reacted to the installation by saying something along the lines of "I can make something like this." Of course, upon closer inspection, they realized that every panel was uniquely and precisely laser cut, printed with color, and digitally labeled for construction. Still, the fact that an

Figure 3.17 The interior of *CHROMAtex* is composed of a spatial gradient of five colors.

everyday object, like the binder clip, gave greater approachability to the structure and deeper insight into the digital nature of its assembly had a large influence on how we began to think about our work.

Reflection: Looking Back at Our Experiments, Prototypes, and Projects, 2014

When we first started investigating the automation of digital production in architecture it was mainly an insular exercise. For example, we were most excited by how digital production processes might produce novel architectural details rather than the cultural or social implication of such processes. We didn't think about how digitally produced details might make a structure feel cold or potentially less approachable than if that same structure was constructed through hand-crafted processes.

Our early interests were also solely driven by conversations with our peers and colleagues in the architectural discipline. As our audience expanded to non-architects, we recognized that we didn't need to demonstrate unerring control over our digital productions—the non-architect crowd was not as preoccupied with digital precision. We thus became increasingly interested in how our work could push back against the common perceptions of digital aesthetics and become more generally accessible. To be clear, we weren't interested in removing, or suppressing, the Digital in our work, but rather in combining digital precision with the inherent idiosyncrasies of everyday details or analog processes. Our driving questions became: how can we construct architecture that is at once automated and precise, but also crafted? And, should machine precision and human craft be readily distinct from one another?

Iteration: Major Leaps Transition to Fine-tuning, 2015–2019

CHROMAtex, and the expanded audience that it helped us build, led to the commissioning of a number of new form-found installations—what began as technological experiments were now being seen as artworks. We initially resisted our work being placed in the "artwork" category; we were and are, after all, designers, not artists. However, we came to realize that calling our research "artwork" provided us with more opportunities for public engagement; as such, we gradually accepted the term.

With more creative opportunities, the development of our automation processes and assembly methods became more iterative and the problems we looked to explore became more nuanced. For instance, as part of an installation at the New Museum's inaugural Festival of Ideas, we integrated batteries and LEDs into the panels in order to make them glow (Figures 3.18 and 3.19). And for installations in the entrance of Etsy's new Brooklyn headquarters and Innisfree's flagship store in Seoul, we experimented with using wood veneer and recycled paper as our building materials (Figures 3.20 and 3.21).

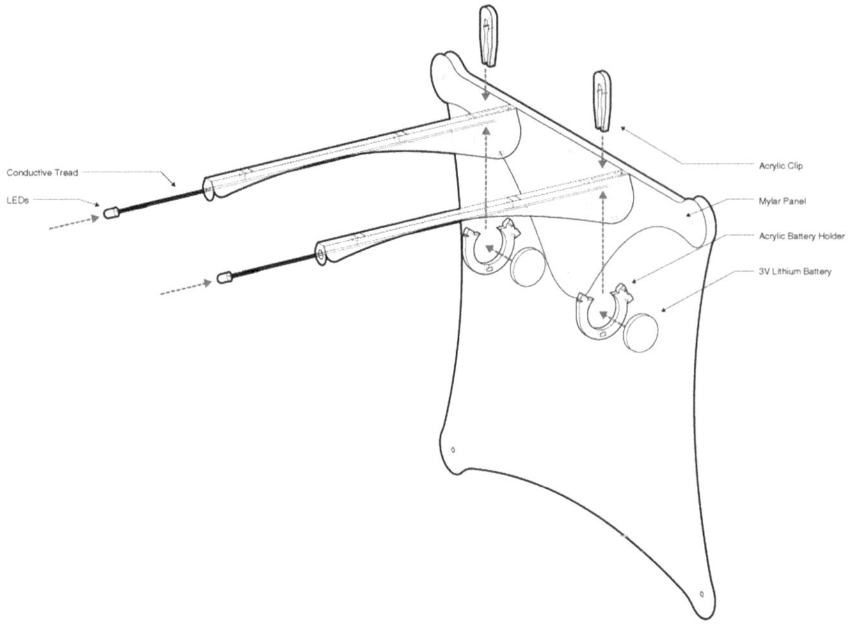

Figure 3.18 Detail drawing showing LEDs powered by watch batteries integrated in a single panel of *POLYPlux*.

Figure 3.19 LED detail of the installation *POLYPlux* created for the New Museum's inaugural Festival of Ideas.

We also created our first prototype using laser-cut aluminum, which enabled us to build more rigid lightweight net structures (Figure 3.22). This prototype became the foundation for an installation at the Southbank Centre in London, in which we created foldable diamond-shaped panels using 3Ms Solar Mirror Film (Figure 3.23). These mirrored panels redirected sunlight as it came through the double-height windows on either side of the Centre's Royal Festival Hall (Figures 3.24 and 3.25). We later used our aluminum fabrication methodologies to create a large form-found installation in collaboration with IBM's AI platform, Watson, for the Mobile World Congress in Barcelona (Figure 3.26). Similar to the Southbank Centre installation, this installation was clad in folded aluminum panels that bounced light in unexpected ways (Figure 3.27).

Figure 3.20 Installation for the main entrance of Etsy's Brooklyn headquarters made with wood veneer and custom-died recycled paper.

Figure 3.21 Installation for Innisfree's remodeled flagship store in Seoul created out of wood veneer and paper made from tangerine rinds.

Figure 3.22 Prototype of a form-found aluminum net structure.

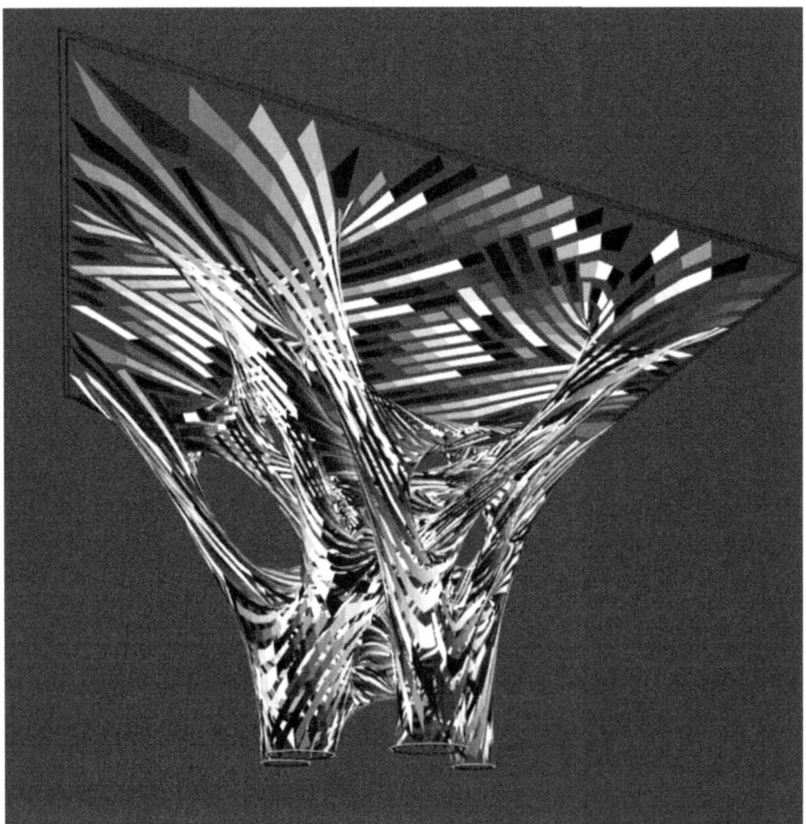

Figure 3.23 Analysis of a form-found surface for directionality of the cladding panels.

Figure 3.24 *Ventricle* installed in the Royal Festival Hall at the Southbank Centre in London.

Figure 3.25 *Ventricle* was made of a rigid aluminum net that was clad in 3M Solar Mirror Film.

Figure 3.26 Installation commissioned by IBM for the 2018 Mobile World Congress in Barcelona.

Figure 3.27 The installation for IBM was clad in folded aluminum panels and 3M dichroic film.

What made this time in our evolution most exciting was that we realized our work could be financially viable. However, once we began to take on more commissioned, less self-driven work, we were inevitably approached by clients asking for projects similar to work they had previously seen us do. In contrast, our earlier work was driven by constantly looking for solutions to pragmatic problems that came so easily to us (whether through disciplinary interests or newly available technologies). We wanted to reintroduce some of the uncertainty present in the early days of the studio, even though we knew it was impossible to forget the knowledge we had gained over the previous decade. How would we recalibrate, or even miscalibrate, the processes we had already developed and refined? We came to realize that finding problems might be just as important as finding solutions.

Idiosyncrasy and Recalibration: Designing Problems as Projects, 2019–present

We subsequently became interested in ways to subvert what we viewed as the overtly rational nature of most digitally fabricated work. For example, we began rethinking the grid-like net structure that undergird most of our previous work and whether such a network could be woven instead. Similar to our binder clip experiment in CHROMAtex, we ultimately wanted to blur the lines between a digitally crafted installation and one driven more by a hand or vernacular craft.

We began to develop design proposals in which we speculated on how we might introduce the idiosyncrasies found in conventional weaving processes into the process of crafting form-found surfaces (Figures 3.28 and 3.29). Alongside these proposals, we began to physically prototype a minimal surface using a system of randomly

Figure 3.28 Model created for a proposal for a public artwork in the entrance of a new children's hospital.

Figure 3.29 Model created for our competition entry for the Chinatown Gateway competition.

Figure 3.30 Initial prototype of an assembly system using randomly placed aluminum strips.

placed aluminum straps. We thought that we could physically recreate the structure of a minimal surface through weaving and that we could stabilize this woven gridshell structure by simply applying random straps to its surface wherever they were needed.

We took a simple catenoid surface that we still often use for prototypes and used a physics simulation tool to digitally stretch random curves around it. We gave each curve thickness, labeled them for assembly, and laser cut them out of aluminum. We were able to achieve the global form, but many of the connection holes between straps did not line up and, as a result, the surface did not naturally take its shape (Figures 3.30 and 3.31).

While this might seem like a failure, this prototype was very exciting for us since we didn't already have a design solution in mind—it felt just like the early days! We made a few more prototypes with random strips but the minimal surface geometry still didn't take its shape. We eventually realized how important the underlying rational grid

Figure 3.31 Detail of the prototype showing that the assembly did not take its natural shape and the parts did not completely line up.

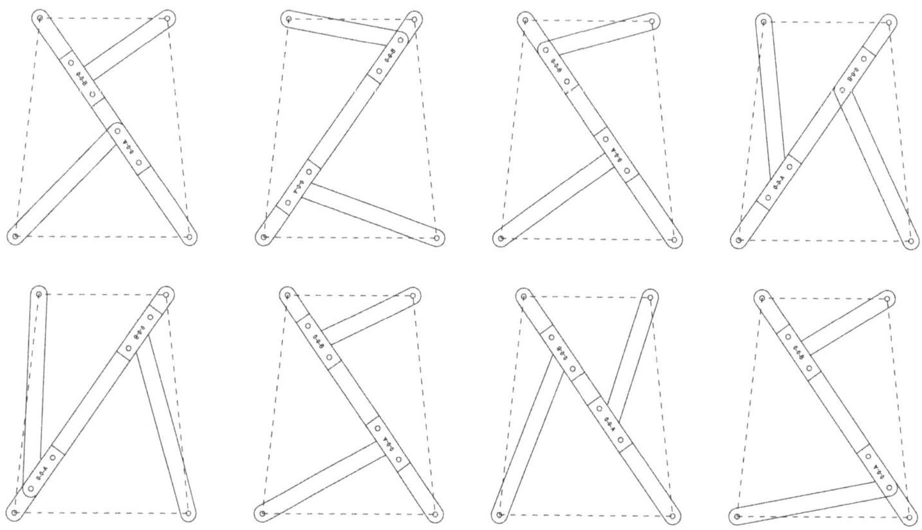

Figure 3.32 Drawings showing the automation of randomly placed aluminum strips that still retain the vertices of the gridshell.

pattern of the surface was for distributing internal forces and defining the global form. This finding was promising because it did not come from our longtime interest in customizing digital tools, but through another avenue that had become integral to the studio: making mockups.

We sketched out a way to maintain the grid points at the corner of each panel but randomize the segments that connected them (Figure 3.32). We laser cut these segments from aluminum (Figure 3.33), and as we assembled, the prototype began to assume its final shape (Figure 3.34). We therefore recognized that maintaining the panel points was the solution. This new prototype achieved the balance between idiosyncratic weaves and rational grids that we were after: the randomized woven segments caused the rational grid to disappear (Figures 3.35 and 3.36). This prototype led to a recently commissioned proposal for the Woman and Children's Hospital in San Antonio, Texas, called "Nest" (Figures 3.32 and 3.37).

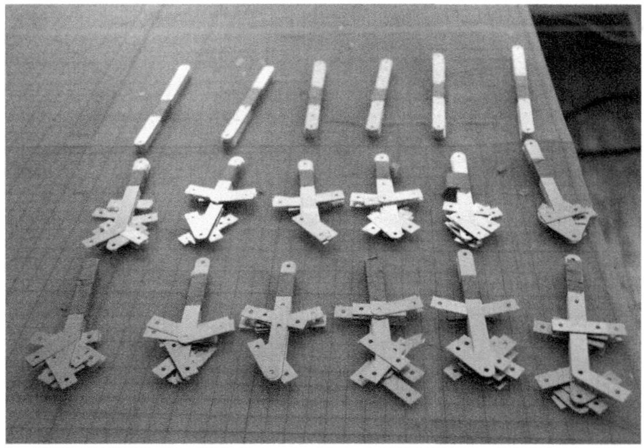

Figure 3.33 Laser-cut aluminum strips organized to assemble the prototype.

Figure 3.34 Photo of the prototype taking its natural shape as it is assembled.

Figure 3.35 Detail of the prototype showing the pattern greeted by the random aluminum strips that retain the rational geometry of the gridshell.

Figure 3.36 Final prototypes to test the aluminum strip width that also confirms the rational gridshell is not visible.

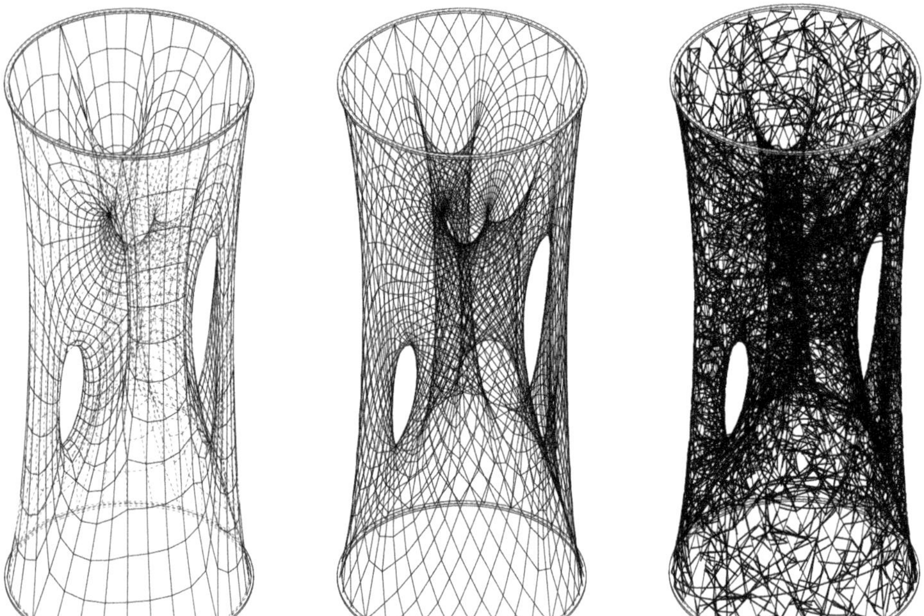

Figure 3.37 Drawings showing the original gridshell, random construction of the gridshell panels, and the layer of randomly placed strips.

This process of rethinking how we might build a gridshell through the use of a non-gridded structure revealed the blurry line between design decisions and design solutions. While the decision to give a gridshell a more woven structure might appear to be a design decision, we knew when we proposed it that it would set up a series of interesting problems for us to solve using form generation and production automation. In the end, this decision not only invited questions on randomness and rationality—two conceptual curiosities of ours—but it also taught us the importance of the grid in the gridshell. This was a great outcome because it led to new conceptual possibilities but also forced us to reevaluate assumptions about the technical knowledge we already possessed.

In the early days of the studio, we believed that solutions drive projects forward, but solutions were not as productive when we started to consider how the studio would work and evolve over time. This latest exploration helped us see that in order to push our work we need to introduce critical problems into our process rather than achieve discrete solutions with our products. However, finding productive problems—ones that cause us to recalibrate the workflows we have already established—is, in many ways, harder than finding interesting solutions.

The woven shell structure we developed for Nest offers us promising insight into what problems might be productive for our practice. Most of the custom processes we have developed over the years, from production automation to form finding to simulation, all rely on rationalization in some way. By introducing ways to subvert that rationalization or by introducing irrational systems, we can do more than simply destabilize our workflows; we can foster new ways to conceive of and test our outputs.

In many ways, we have been introducing productive problems all along. But it has taken naive exuberance, getting lost, introspection, and many, many mistakes for us to view problems as a productive, and necessary, part of how our studio and our work develops. The introduction of problems, whether they are born from attempts at automation, mapping gradients, or weaving laser-cut aluminum, can lead to new creative directions and ways of evaluating how the studio approaches both commissioned work and research.

Notes

1 Autdodesk's Maya keeps track of geometric transformations and modeling through a node based construction history that allows for the live editing of steps used in the modeling process.

2 The name is based on Lewis Carrol's beloved fiction, *Through the Looking-Glass*. In the story, the name, "Alice," is coopted as a term in physics to describe a mirrored topology.

Not *Not* Digital Fabrication Machines

Maria Yablonina

Introduction

A few years ago, I submitted a paper to one of our field's shinier conferences in which I described my practice of designing machines as an architectural method.[1] In the paper, I discussed a project that demonstrated the way I tend to work: discover a site – design machines to match it – choreograph the machines to fabricate an architectural object in situ. Like many of my projects, this one concluded with a set of small robots crawling up and down the surfaces of a room, fabricating a woven structure made of brightly coloured rope (Figure 4.1). I staged this approach as a site-specific alternative to using industrial robotic fabrication equipment. Reviewer 2 responded: "*this is an interesting design idea but one that misses the main asset of industrial robotics–versatility. An industrial robot can do many things; this system will do just one.*" Years later, I am still wondering if I did indeed miss something in my struggle to appreciate the value of industrial fabrication equipment's so-called "versatility." I remain skeptical. I am convinced that alternative, non-versatile machines have a place in design and fabrication—adopting industrial robots that "can do many things," along with the neoliberal values embedded within them, can't be the *only* way to move forward.

 DOI: 10.4324/9781003296522-5

Figure 4.1 Mobile Robotic Fabrication System for Tensile Filament Structures 03. A multi-robot fabrication system for in-situ woven structures. Documentation of the installation and troubleshooting process. Kapelica Galley, Ljubljana, Slovenia. 2019.

Origin Story

I became a machine maker by accident. As a technology-obsessed graduate student in architecture, I was fortunate to begin my academic career at a time when industrial robotic tools were just beginning to appear in architecture schools, and computational thinking[2] promised a bright future for the profession. The potential of using robots in design was new, unexplored, and thrilling. Moreover, I was privileged to be enrolled in one of the few architecture programs that not only had a robot arm but also provided an exceptional level of in-house expertise in operating it. I was beyond excited to use robots in design, but never planned to have a career in developing any myself.

In the final year of my graduate studies, my class had access to a single robot arm. Split between 16 students, machine time was coveted, to say the least.[3] Despite my deep infatuation with the arm, I decided that constructing my research without relying on this singular resource would be more convenient and, ultimately, more productive. I therefore resolved to construct my own fabrication equipment that would be mine alone. In this ironically unpragmatic and naive determination, I stumbled into a technical design practice that forced me to re-imagine my relationship to machines, their relationship to the built environment, and ultimately my understanding of what a design practice can be.

The Double Negative

In his essay "Anti-anti: Utopia, Globalization, Jameson," Eric Cazdyn describes the double negative as a "stylistic and rhetorical tactic" for finding new possibilities between impossibly flawed positions.[4] In Cazdyn's description, the double negative is neither an affirmation nor a rejection of thought but "implies a special temporality" in relation to it. I wasn't aware of Cazdyn's double negative when I first started designing my machine prototypes, but in making them I have been intuitively employing the anti-anti tactic all along.

The systems I have built since my initial move into machine-making embody the double negative tactic in more than one way: my machines do not move in the same way as industrial equipment, are not constrained by the same limitations, do not meet the same performance criteria, and are not carrying the same metaphors. And yet they are not *not* digital fabrication machines. In this tension of simultaneous familiarity and departure, my machines offer a theoretical and technical position from which to imagine alternatives and critique the status quo of robotics in design and architecture.

Not *Not* Versatile

Like any other tool, digital fabrication machines afford opportunities at the cost of limitations; for instance, the work envelope and the payload of a robot arm define the scale of objects that can be produced, the size, weight, and power of an arm inhibit its deployability in unstructured spaces, and the cost, complexity of installation, and steep learning curve limit the machine's operation to expert users in expertly equipped spaces. Intimate familiarity with constraints is part of understanding any medium, and digital fabrication technology is no different. In developing this instrumental relationship, we build an intuition with our tools: we learn their quirks, grasp their limitations, and come to sense where and when to push them beyond their limits.

Over the past two decades, the community of researchers in computation and digital fabrication in architecture has developed a remarkable understanding of its tools. We have developed methods for overcoming and compensating for the limitations that the machines have inherited from their past factory roles. We have figured out how to harness their precision, repeatability, and strength in a myriad of materials, forms, and fabrication sequences. However, I fear that in working with production machines, we have elected to bend architectural practice and its object to fit the industrial equipment and not vice versa. Unsurprisingly, a machine designed for serial manufacturing of car parts can do many incredible things, as long as it remains in a factory-like scenario making car-like objects. As a result, so much of the contemporary research in robotics in architecture seems to be focusing on the fabrication of new forms with new materials and their assemblies into new architectures. But what of the architectures that are already here?

Back when I first decided to make my own machines, I had to figure out where to begin. Starting a digital fabrication project with no fabrication machine at all produces anxiety like that of writing the first sentence of an essay: the possibilities are endless, the potential is overwhelming, and disappointment in the outcome feels imminent. I

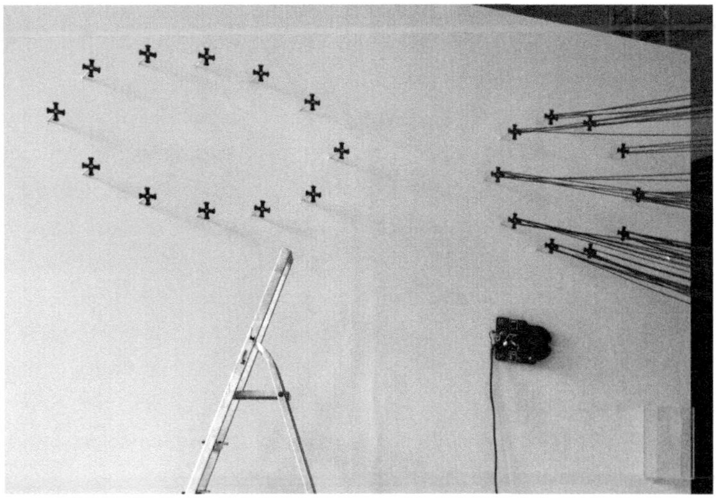

Figure 4.2 Mobile Robotic Fabrication System for Tensile Filament Structures 01. A heterogeneous multi-robot system for winding of filament structures in situ. The system consisted of two wall-climbing and one cable-walking robots cooperatively winding a sculptural object using rope. Documentation of the fabrication process and setup. The ladder did not make it into any of the official documentation documents, despite its essential role as the troubleshooting tool. 2016. Milan. Italy.

urgently needed some constraints and criteria. Whether by intuition or habit, I found them in a familiarly architectural place: site, scale, material. I began by searching for ways to occupy spaces that industrial fabrication equipment could not reach. I chose to prioritise sites that were mundane and elements that were commonly present in the everyday architectural contexts: walls, ceilings, beams, and columns. I ended up developing systems that use building stock as a substrate for new layers of architectural tasks (Figure 4.2). In designing machines for these substrates, I found an opportunity to subvert the logic of robot-oriented design, i.e., the notion that the designed object must fit within the machine's fabricability constraints.[5] Instead, my machines were the ones that had to fit in the narrow overlap, both physical and conceptual, between the surface of an interior wall and the boundary of the fabricated object.[6]

The machines I have been making are always task-specific: just fitting the design criteria demanded by their sites, not a task more. Abandoning versatility is what made this approach feasible: the machines could be simple and yet functional in the one task they were expected to perform. By sacrificing the versatility of tasks, I gained a multiplicity of potential sites. For instance, I developed a wall climbing machine that can be deployed on walls in most conventional interior spaces, as long as they meet the minimum criteria of surface finish.[7] I have positioned these machines as contributions to the existing catalogue of tools that promised greater access and an array of new sites for digital fabrication. I later called this approach *Designing [with] Machines* and described it as a collection of interdependent relationships: machine-environment, machine-material, machine-machine, and machine-designer.[8] Of course, these emerging relationships are not linear, and the hierarchy of their parameters is messy. Despite my attempts to classify and describe the machine-site-material parameters, they were difficult to separate, as they continued to stubbornly seep into one another. Each new project diagram felt like an entangled mess of values: site geometry, program, motor torques, work envelopes, object geometry, scale, structure, material, aesthetics, time, cost... Every time a parameter within one of the relationship pairs had to be adjusted, the change propagated throughout most variables, in often unexpected and destructive ways (Figure 4.2). For instance, with a rope-weaving robot, a change in rope material and/or thickness required a change in locomotion motor torque to accommodate new material weight. A change in motor torque necessitated a change in motor size, which in turn impacted the size of the machine's footprint. A larger machine was less manoeuvrable, meaning that both path-planning and hook-placement strategies had to be adjusted. Hook placement dictated the resolution and pattern possibilities of the wound object, which in turn propagated all the way back to the choice of rope material.

Not *Not* Research Methods

Caught inside an often-paralysing indeterminacy of intertwining parameters, I found myself continuously mediating between many agendas: my own, the site's, the material's, and the machine's. In this mediation I had to build an understanding of each, gradually learning which boundaries I could push, and which were better left untouched. I have not quite noticed when it happened, but the whole process became a conversation:

ME:	Could you climb walls?
MACHINE:	Maybe...
ME:	Amazing! How high do you think you could climb?
MACHINE:	Emm... Depends on the wall. Is it flat? Is it smooth?
ME:	I don't know yet... I guess let's start with this wall right here in my living room? Could you climb that?
MACHINE:	Are you sure you can afford to make me battery powered? I assumed that was way outside your budget and schedule...
ME:	Right... Are you saying you can climb as high as the power cable is long?
MACHINE:	Yeah! Until the cable is so long that it is too heavy for me to lift.
ME:	I guess I'm going to have to redesign all the fabrication sequences to make sure the cable doesn't get caught on things?
MACHINE:	I guess so...

Leading up to the final iterations, calibrated fabrication choreographies, and polished documentation, the systems went through many iterations of actively falling apart (Figure 4.3). Machines that may have worked perfectly in lab tests disintegrated into useless parts once brought to real world sites. Even in their finalised states, they were often flimsy, finicky, temporary, and hard to operate (Figure 4.4). Despite me raising them to be architectural fabrication robots, they consistently strove to become Minsky's useless machines.[9] Instead

Figure 4.3 Mobile Robotic Fabrication System for Tensile Filament Structures 02. A multi-robot system for winding of filament structures using suspended sheet elements as substrate for robotic locomotion. Documentation of the troubleshooting process. 2016. Ars Electronica. Linz, Austria.

Figure 4.4 Mobile Robotic Fabrication System for Tensile Filament Structures 03. 2019. Documentation of the troubleshooting and setup processes.

of seamlessly integrating into the design process and the architectural environment, they demanded attention: never quite ready-to-hand, but stubbornly present.[10]

Refusing to disappear into their tool-ness, my machines demanded that I take time and space to contemplate their function and context; they became research aides rather than merely fabrication tools. Watching them reluctantly navigate the surfaces of their environments, I began to notice the stark contrast between the utopian stories we tell about technology and the reality of architectural spaces we already occupy. While prototyping, setting up, and calibrating these task-specific machine systems in built environments, I got to spend a lot of time in proximity to unevenly painted textures, non-planar walls, cracks in materials, and seams between elements. Developing controllable and predictable systems for operation in these utterly incomputable environments demanded that I accounted for it by building in tolerances, compliancy, and responsiveness. Sometimes, it simply demanded that I let go of precision and allow the systems to be flexible. Flimsiness became a design criterion as I moved on from one project to another. I learned to anticipate the unexpected and plan accordingly for adaptation rather than control. Every new system became less precise and, as a result, a little more successful in what it could accomplish, revelling in its imperfection.

Not *Not* Political Artifacts

Over time, I built an intimate understanding of flimsiness. I developed an intuition with my tools: I understand their quirks, feel their limitations, and know when they need me to take over. I learned to design and build systems with a human in mind: not to

Figure 4.5 Mobile Robotic Fabrication System for Tensile Filament Systems 03. 2019. Documentation of the fabrication process.

automate, but to augment, to design myself into the system as one of the fabrication team members (Figure 4.5). I feel deeply connected to my machines and to the spaces that they occupy. As this relationship developed, my infatuation with the narrative of versatile machines and autonomous robotic fabrication in architecture has faded.

In the conclusion to "Anti-anti: Utopia, Globalization, Jameson," Cazdyn writes: "the double negative might be called a phrasing that formally captures the dialectic of identity and difference."[11] In this seemingly contradictory position, Cazdyn describes a tool that simultaneously questions status-quo methodologies and "is reserving its toughest criticism" for the politics and cultural contexts in which those methodologies have emerged. In their not not digital fabrication-ness, my machines perform a similar critique. They do not oppose or aim to replace industrial digital fabrication equipment. In fact, arguing for such replacement would fit within and further perpetuate the narrative of one tool's superiority over others. Instead, my machines attempt a lateral move towards a multiplicity of methods – they offer an opportunity to dissect the prominent robotic fabrication narrative from a newly gained distance. In probing the validity of the versatile robot, my machines point to a bigger picture: the political implications of versatility as a design metric.

In his essay "Do Artifacts have Politics," philosopher Langdon Winner argues that any technology contains the politics of the context, history, and economy it originated from.[12] Winner suggests that technical methods and tools imply (intentionally or not) political metaphors in how their bodies, systems, networks, and infrastructures are arranged. Following Winner's urge to look for the political in the technical, I see digital fabrication tools in a new light. In their exceptional precision, versatility, and efficiency, contemporary fabrication machines resemble their predecessors: industrial labour automation equipment powered by the post-war restructuring of military research resources. I wonder how much this lineage has contributed to a prominent discourse that seems to increasingly value efficiency and automation above all else (Figure 4.6).

My flimsy machines have taught me that the technological futures we imagine in our discourse cannot be as seamless and shiny as the labs where most of the imagining occurs. These futures will have to be built upon and integrated within existing contexts of messy, imprecise, and mundane architectural environments: ones that demand care, attention, and repair rather than demolition and re-building. If we are serious about constructing responsible architectural futures, we must learn to occupy imperfect spaces. Such a shift requires that we take care and adapt, question our priorities, and extend our focus beyond fabrication of beautiful new objects toward maintaining and augmenting what is already in place: the unglamorous and often un-computable spaces of the existing building stock. One way to approach this shift is to prioritise repair over construction and labour augmentation over automation.[13] I believe that this shift requires not one but a myriad of anti-anti digital fabrication approaches: ones that simultaneously acknowledge the powerful potential of tools and are critical of technology's embedded politics.

Figure 4.6 Mobile Robotic Fabrication System for Tensile Filament Structures 01. A heterogeneous multi-robot system for winding of filament structures in situ. Documentation of exhibition opening performance. Originally the system was intended to run autonomously, but due to direct sunlight interference with localisation system had to be operated manually throughout the performance.

Not *Not* Conclusion

I do not urge the reader to walk away from their equipment in pursuit of better politics. Nor do I suggest that the machines I have developed can assist in such a pursuit. The systems I have developed are unmistakably mine: they are a product of my design choices, my expertise, my budget limitations, my time, even my social circles, support structures, and values. They are in no way neutral or free of metaphor. They carry my biases, speak on my behalf, and therefore inevitably signal my politics, whether I intend them to or not (Figure 4.7).

To follow Winner's provocation to its logical conclusion, I must acknowledge further resolutions of metaphor embedded in the tools. My machines are also a product of their literal parts: motors, sensors, bolts, nuts, rubber gaskets, tape, vacuum cleaner parts, 3D printing filament, CAD software, microcontrollers, and many others whose origin stories they carry within them. Created and placed in the discourse of digital fabrication in architecture they are inseparable from its context and metaphors, too. Existing at this intersection of many contradicting agendas, I can only hope that my machines

Figure 4.7 Mobile Robotic Fabrication System for Tensile Filament Structures 01. Documentation of the wall-climbing robot assembly process.

perform my favourite function, which is to point loudly at the contradictions within the technological conventions and make space for contemplating myriad double negative versions of computational design thinking.

Notes

1 Maria Yablonina, Nicholas Kubail Kalousdian, and Achim Menges, 'Designing [with] Machines: Task- and Site- Specific Robotic Teams for in-Situ Architectural Making', in *ACADIA 2020: Distributed Proximities. Proceedings of the 40th Annual Conference of the Association for Computer Aided Design in Architecture*, eds. Viola Ago et al., 2021.

2 Here, it is crucial to think with two versions of the definition of "computational thinking". One, having been central to the field of digital fabrication, is offered by Achim Menges and Sean Alquist: in opposition to computerisation, computational design thinking relies on algorithmically processing information and producing output at the intersection of many interdependent inputs. See Achim Menges and Sean Ahlquist, eds., *Computational Design Thinking*, AD Reader (Chichester: Wiley, 2011), 11. The other definition, offered by James Bridle establishes computational thinking as "*the belief that any given problem can be solved by the application of computation [...] computational thinking supposes [...] that it is impossible to think or articulate the world in terms that are not computable*". See James Bridle, *New Dark Age: Technology, Knowledge and the End of the Future* (London; Brooklyn, NY: Verso, 2018), 14.

3 This happened just a couple of years prior to Antoine Picon writing about robot arms as status symbols: indicators of wealth and power wielded by a design institution that can afford them. See Antoine Picon, 'Robots and Architecture: Experiments, Fiction, Epistemology', *Architectural Design* 84, no. 3 (May 2014): 54–59, https://doi.org/10.1002/ad.1754. I suspect that witnessing this tension between perceived status and the reality of coveted machine time has contributed to my discomfort with the inherent exclusivity of power that these machines grant.

4 Eric Cazdyn, 'Anti-Anti: Utopia, Globalization, Jameson', *Modern Language Quarterly* 68, no. 2 (1 June 2007): 331–343, https://doi.org/10.1215/00267929-2006-041.

5 Thomas Bock and Thomas Linner, *Robot-Oriented Design: Design and Management Tools for the Deployment of Automation and Robotics in Construction* (Cambridge: Cambridge University Press, 2015), https://doi.org/10.1017/CBO9781139924146.

6 Besides the architectural constraints of site, design, and program, these machines also had to squeeze between the borders of spreadsheet columns that separated my planned budget and the minimum

necessary hardware that was required to construct them. Moreover, they had to fit within the constraints of my slowly developing mechanical and electrical engineering skills (or the lack thereof).

7 The wall climbing machines that I describe in this section were capable of traversing most conventional interior surface finishes: paint, wallpaper, etc. Some hardware calibration was required to traverse rougher surfaces of exterior wall finishes or glass window surfaces.

8 Yablonina, Kubail Kalousdian, and Menges, 'Designing [with] Machines: Task- and Site- Specific Robotic Teams for in-Situ Architectural Making'.

9 Arthur C. Clarke describes Minsky and Shannon's machine: "Nothing could look simpler. It is merely a small wooden casket the size and shape of a cigar box, with a single switch on one face. When you throw the switch, there is an angry, purposeful buzzing. The lid slowly rises, and from beneath it emerges a hand. The hand reaches down, turns the switch off, and retreats into the box. With the finality of a closing coffin, the lid snaps shut, the buzzing ceases, and peace reigns once more <…> There is something unspeakably sinister about a machine that does nothing—absolutely nothing—except switch itself off." See Arthur C. Clarke, *Voice Across the Sea* (New York: Harper & Row, 1959), 159.

10 My relationship to Heidegger's notions of tools being either "ready-to-hand" or "present-at-hand" is complicated. I acknowledge that the terms are often overused, and their invocation inevitability, albeit unintentionally, invokes the politics of their author. And yet, denying the importance of Heidegger's writing to my ways of thinking about and relating to machines would have been disingenuous. See Martin Heidegger, *Being and Time*, Reprint (Malden, MA [u.a.]: Blackwell, 2002).

11 Cazdyn, 'Anti-Anti'.

12 Langdon Winner, 'Do Artifacts Have Politics?', *Daedalus* 109, no. 1 (1980): 121–136.

13 Maria Yablonina and James Coleman, 'Small Robots and Big Projects', in *Realignments: Proceedings of the 41st Annual Conference of the Association of Computer Aided Design in Architecture (ACADIA)*, 2021.

Bibliography

Achim Menges and Sean Ahlquist, eds., *Computational Design Thinking*, AD Reader. Chichester: Wiley, 2011.

Arthur C. Clarke, *Voice Across the Sea*. New York: Harper & Row, 1959.

Eric Cazdyn, 'Anti-Anti: Utopia, Globalization, Jameson', *Modern Language Quarterly* 68, no. 2 (1 June 2007): 331–343, https://doi.org/10.1215/00267929-2006-041.

James Bridle, *New Dark Age: Technology, Knowledge and the End of the Future*. London; Brooklyn, NY: Verso, 2018.

Langdon Winner, 'Do Artifacts Have Politics?', *Daedalus* 109, no. 1 (1980): 121–136.

Maria Yablonina, Nicholas Kubail Kalousdian, and Achim Menges, 'Designing [with] Machines: Task- and Site-Specific Robotic Teams for in-Situ Architectural Making', in *ACADIA 2020: Distributed Proximities. Proceedings of the 40th Annual Conference of the Association for Computer Aided Design in Architecture*, eds. Viola Ago et al., 2021.

Maria Yablonina and James Coleman, 'Small Robots and Big Projects', in *Realignments: Proceedings of the 41st Annual Conference of the Association of Computer Aided Design in Architecture (ACADIA)*, 2021.

Martin Heidegger, *Being and Time*, Reprint. Malden, MA [u.a.]: Blackwell, 2002.

Thomas Bock and Thomas Linner, *Robot-Oriented Design: Design and Management Tools for the Deployment of Automation and Robotics in Construction*. Cambridge: Cambridge University Press, 2015.

Beyond Optimization

Interaction as a Means to Resituate Digital Fabrication

Stefana Parascho

Introduction

The field of digital fabrication in architecture seems to be slowly approaching a point of saturation in which the once seemingly endless range of technological possibilities will be flattened into a singular, technopositivist pursuit. In my opinion, such flattening is not a symptom of the field already having run its course, but rather a consequence of several overlapping factors, including the confused expectations of many digital fabrication researchers, such as myself, that practice at the intersection of design and the hard sciences; the ways in which we tend to privilege quantifiable validation over other metrics; and the large amounts of time we often spend solving technical problems rather than formulating them.

In this case study, I will attempt to unpack how the field of digital fabrication has manoeuvred itself into a corner of prescriptive processes and predictive evolution. I aim to do this by analysing how the most popular aim of digital fabrication research, optimization, is used in both my own work and the work of others in the field. In particular, I will examine how our collective fascination with optimization has led us towards a top-down approach to working with machines and away from more creative, improvisational design processes. What follows is not meant to be an objective analysis of recent developments in digital fabrication, nor is it a comprehensive exposition of the challenges currently facing the field. Instead, I try to relay my own understanding of architecture-oriented digital fabrication as an active participant in the field for 15 years and to reflect upon my practice in

DOI: 10.4324/9781003296522-6

order to recalibrate its goals, methods, and possibilities. This case study comes at a critical moment in my own academic development in which years of overtly technical research have caused me to question how the framing of my work influences the directions it can, or *should*, take. My hope is that by critically reflecting on my own digital fabrication work and laying out new intuitive directions for exploration, other researchers may feel inspired to look for their own ways out of the dominant, technical narratives of the field and into more open-ended, even risky, endeavours.

I would argue that the field of digital fabrication has organically evolved with little-to-no collective agreement on what type of research questions should guide its development or how research should be evaluated. When it first arrived on the architectural scene at the end of the 20th century, digital fabrication provided designers with justification for the exploration of new formal possibilities; at the same time, it built upon the development of the computer and other physical machines that were invented by engineering disciplines in decades prior. This duality—between artistic pursuit and mechanical invention—initially led to a variety of nuanced digital fabrication research questions and approaches that allowed researchers to exploit the field's inherent in-betweenness.

Today, this variety continues to diminish in favour of prescriptive research approaches that seem to centre on efficiency and optimization. For example, we frequently make use of computational design and digital fabrication tools to generate geometries that optimize material use, time, cost, etc. More often than not, such optimization is purely an exercise in maximization or minimization—we generally don't ask ourselves what might constitute an *optimum* for a specific project beyond obvious quantifiable measures.

Optimization is repeatedly used to create a seemingly irrefutable argument for the use of digital design and construction methods. While the most popular goals of optimization do address critical, global issues, such as natural resource depletion, an often overlooked side-effect of these technical efforts is that they promote and institute top-down processes that define goals, methods, and tools in inflexible ways. For instance, once optimization targets are established, research is often generated and tested on repeat until those targets are reached; there is little space left for evolving the project's questioning and rethinking its goals once the optimization process is set in motion. Further, the definition of optimization goals often excludes other qualitative factors that should be considered in an architectural design project, such as its physical, social, or economic context and, I would argue most important of all, human interaction.

In her book *Undoing Optimization*, the social theorist Alison Powell argues, "the big data optimized smart-city vision places value on the generation of data and its calculation, curation, and aggregation, making explicit assumptions that these activities, rather than the generation of ideas of knowledge, are what matters."[1] While Powell's discussion centres on questions of smart cities, big data, and surveillance, she raises issues that are germane to conversations on robotic construction, including the ways in which participation and interaction are typically suppressed by the increasing use of current digital technologies. Powell suggests that by focusing on optimization, we are devaluing the role of creativity and any new ideas that don't directly contribute to the solution of a given optimization problem. Powell concludes her argument with a call for more technological openness and flexibility through a hybrid approach to technology that simultaneously makes use of data and computation and allows for human participation, she writes, "We should be able to retain rights and capacities to hybridize – to transform and evolve our ways of knowing as the world we inhabit continuously becomes less certain and less easy to narrow and optimize."

In the field of digital fabrication, the lack of hybridization, or willingness to hybridize, has led to the prevalence of rigid top-down approaches. Besides taking fabrication out of context (it is generally assumed that any design could be fabricated anywhere in a similar manner), these approaches typically exclude humans from the creative loop. Working with top-down fabrication processes often leads to forms that merely respond to optimized parameters, without opening space for critical questioning or experimentation, especially during the fabrication phase.

I see digital fabrication's seemingly irreversible consensus on the goals of optimization and efficiency as a symptom of a confused field, still searching for its place in the architectural discipline, and the academic world at large, at a time of colliding global social, health and environmental crises. Our current global crises make a strong case for applied research that has an immediate, quantifiable impact. But this drive for immediate action has compelled our relatively new subdiscipline to quickly decide where it positions itself: are we in a creative field open to new ideas or in a close-loop focused on problem-solving? Maybe what we need at this moment in time is not more technical problem-solving, but to rethink our own relationship to technology.

The Role of Failure and Error: Deviation from the Goal

The field of digital fabrication is governed by a culture of success. Coupled with our fixation on success, and accessory to it, is our simultaneous reluctance to appreciate, or even admit, failure. We generally only set goals at the beginning of research projects and dedicate the majority of our efforts to finding ways to achieve these goals with the utmost precision—not reaching these goals is not an acceptable outcome and changing them along the way is often perceived as a marker of failure. To be fair, admitting failure is something rarely seen in academic research in the field. Digital fabrication research seems to have created its own particular brand of success, merging the expectations of glossy architecture images with those of technical research by presenting mostly polished research results that exclusively showcase feats of precision and efficiency.

As architectural theorist, Francesca Hughes describes in her book *The Architecture of Error*, "All of the architect's anxieties about not being 'scientific', not being the one that actually builds the building and never quite being in full control of material behaviour, meet in architecture's construction of precision." Hughes is purposely separating error from failure, referring specifically to material error, as she beautifully points to the discomfort that error and imprecision evoke within architects, and how digitalization is aggravating this discomfort by heightening the expectation of absolute accuracy. If we, as architects, cannot accept error, then how can we, as digital fabrication researchers, expect failure, adaptability, and indeterminacy to be parts of our practices?

In my own research, I want to ask: can we use errors and failures to create spaces for reflection, adaptation, and analysis? What methods or evaluative criteria would we need to invent to facilitate such a rethinking? And finally, can we, digital fabrication researchers, overcome institutionalized notions of "research excellence" and renew our shared definition of success by revaluating error in our research methodologies?

The Roles of Humans and Machines

In top-down digital fabrication processes, the roles of all actors involved are rigidly defined and typically inflexible. I have chosen to use the term "actors" here in order to question the dichotomy between humans and machines and leave their separate roles more open to interpretation. If the aim of a digital fabrication research project is solely to solve a quantifiable problem, most steps or tasks in that project will be determined before any design or fabrication work begins. Digital fabrication machines, thus, often merely follow predefined instructions. In many cases, such determinacy also means that there is no information flowing in the opposite direction, that is, from machine to design or from machine to human operator. Of course, one can point to examples in which sensing is used to measure tolerances or document unexpected changes in the design environment. But the goal in such instances is typically to identify if the built object deviates from the initially planned geometry and, if so, to find ways to increase fidelity between what is planned and what is built. A typical robotic fabrication research workflow might thus follow this structure:

1. Design problem and optimization goals are defined.
2. Computational tools are developed based on the goals defined.
3. An "optimal" design is generated from the computational tools.
4. Robotic instructions are generated from the optimal design.
5. A structure is fabricated by executing the robotic instructions, with some space for limited feedback and adaptability.

The goal of such a workflow is generally to reduce the human intervention required or to explicitly separate the role of the human from that of the robot. But how many research projects that utilize these kinds of workflows actually go as planned? Can or should we predefine every single step in the fabrication process and expect each one to work out without incident? We could indeed put a lot of time and effort into trying to predict everything that might diverge from our original work plans. But construction and architecture are, at least in practice, highly unpredictable processes with bespoke parts and little repeatability: trying to foresee every possibility of failure, challenge, and change would be extremely tedious, if not impossible. At the same time, as I have argued earlier, error and failure are not necessarily detrimental, especially when mitigating design and construction complexity. There is value in leaving the final, fabricated result more open, and in pairing error with reflection to evolve design and construction alongside other design parameters such as site conditions, social and economic context, information flow, etc.

Instead of defining particular outcomes and the precise methods that may enable us to achieve them, I'd argue that we can learn from on-the-fly design and fabrication processes that allow us to interact with the physical, cultural, and social environments in which construction occurs. In other words, I want to ask whether we can produce more innovative results if we simply give both ourselves and our machines more time to explore, learn, and adapt.

We see such approaches to innovation in other creative fields, such as the visual arts. In such projects, the relationship between humans and machines is often questioned, tested, and reinvented through human-machine interaction. Sougwen Chung's work, for example, offers insights into the ways in which the roles of machines can constantly change throughout processes of creation.[2] By co-creating with machines and allowing each actor, human and machine, to learn from the other, Chung generates art pieces that *embody* interaction (Figure 5.1). Madeline Gannon's performances[3] similarly demonstrate a more emotional connection between humans and machines than those we typically build with robots in our own field. Gannon develops interactive control methods that allow machines to directly react to human input; her workflows display the potential benefits of humanizing objects and machines (Figure 5.2).

While the applications of these artistic projects are not solution-oriented, as they are in the technically oriented discipline of digital fabrication, they reveal another dimension of human-machine interaction that I believe can transform our own creative processes. In these projects, makers engage with machines through emotional response, reaction, and reflection rather than by merely instructing them. Such a human-machine dialogue can help designers to navigate the complex, and oftentimes unquantifiable, contextual relationships that exist in a construction project.

I now want to take a closer look at robotic fabrication processes. In particular, I want to focus on the roles of the individual actors involved in these processes, both human and robot, in order to analyse how the two approaches described above—top-down and open-ended—can be implemented in a design and robotic assembly problem.

Figure 5.1 Collaborative painting. *Drawing Operations,* Sougwen Chung, 2017.

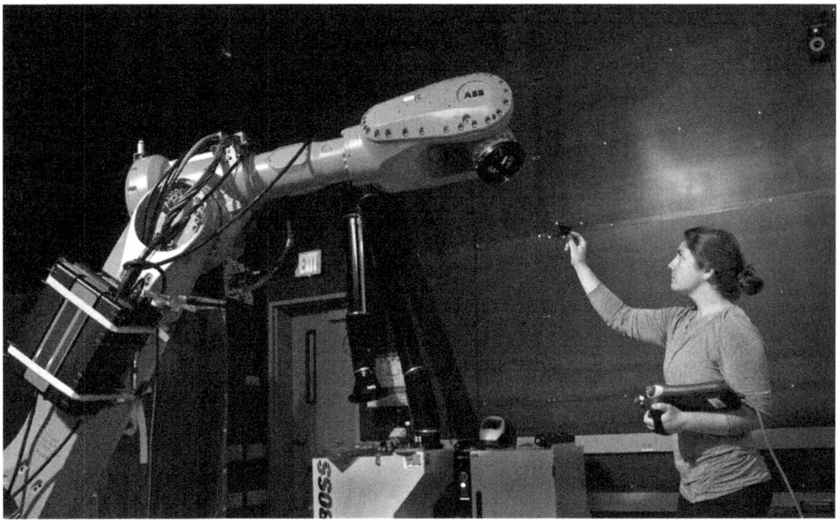

Figure 5.2 Human-robot interaction. *Quipt*, Madeline Gannon, 2017.

My goal here is to show that while many digital fabrication processes attempt to keep the roles of human designers and robotic fabricators separate, humans often still become spontaneously involved in the fabrication process, if only as a last resort to solve unpredicted problems. Such spontaneous interactions suggest that involving humans in an improvised, unplanned manner in an automated process can be productive. Further, I want to show that in order to invite humans into digital fabrication processes, we need to rethink the inflexible optimization-centric objectives that guide most of our methodologies and evaluative criteria.

LightVault

The LightVault was a collaboration between myself, the Formfinding Lab at Princeton University, and the architecture and engineering firm, Skidmore, Owings, and Merrill (SOM) (Figure 5.3). The project centred on a common conceit of construction robotics: the idea that robots can do certain tasks better than humans. In this particular case, such tasks included precisely positioning bricks in space and providing steady support for complex brick assemblies. Based on these perceived advantages, the aim of the project was to allow for the scaffold-free construction of doubly-curved brick structures that could optimize material usage and structural stability.[4] This was specifically achieved by using two 6-axis industrial robotic arms cooperatively to alternate the role of positioning and supporting.

The project's clearly defined objectives produced clearly defined roles for both robot and human actors: the robots alternated the role of construction and support; the humans, in turn, solely engaged in the fabrication process by manually adding the adhesive between bricks as the robots held them in place. Such a fabrication workflow exhibits a logical distribution of tasks: robots' strengths include precise positioning and the

Figure 5.3 ***LightVault,*** **CREATE Lab Princeton and SOM, 2020.**

capacity to both hold and support a structure without moving for long periods of time; humans, on the other hand, are good at tasks that require more dexterity, such as inserting glue in gaps between the bricks. The project's division between design and construction was also clearly defined: the human generated the design using computational form-finding methods and analysed it for structural stability using simulation software; the robots' only purpose was to build.

In reality, things worked only partly as described above. It is in this discrepancy between the project's idealization and execution that I begin to identify research opportunities that we, as a field, have not yet extensively explored. Several unpredicted problems appeared during fabrication that had to be addressed through human interventions and on-the-fly adjustments. First, we only physically tested a few critical tool paths that we knew would present challenges due to the robot configuration before starting construction; most other movements were not verified. This minimal calibration led to many instances in which the robot ran into singularities or reached an axis limit. The solution to such errors was relatively simple. The human operating the robot would either adjust the robot's configuration manually to ease the insertion of the brick or, if this was not feasible, would plan an alternative path by interpolating between a series of points the robot could reach. Since the reachability of bricks was estimated based on the robot's reach, and the most extreme cases of reach were tested, there were no situations in which a brick's final position was completely inaccessible.

Another difficulty we observed during construction was related to tolerances. The robotic setup had limited absolute accuracy due to the robots' installation and the mechanical inaccuracies of the machines. As a result, the offset between the two robots measured up to 5 cm between their two most extreme positions. This was also solved through human intervention, in this case, by testing the offset between the two robots at key points in the fabrication process and adjusting the 3D model accordingly during fabrication.

Perhaps the most dramatic challenge we faced during the project was the requirement to stop construction sooner than anticipated due to the COVID-19 pandemic. As a result, we adjusted the design mid-construction to consist of a smaller number of bricks than we initially planned for so that the vault could still be completed on time. The ability to make these changes in real-time, which was afforded in part by the digital design and fabrication technologies we used, allowed us to successfully construct the vault, even under uncertain conditions (Figure 5.4).

Of course, some of these challenges were predictable and simply the result of a limited timeframe between planning and production; our ad hoc troubleshooting was thus somewhat to be expected. Still, we were able to save, or optimize, time and planning by allowing human actors to directly intervene in the process, even with small changes. Such improvisation provoked a number of questions: what if we purposely allowed more flexibility in the design definition, task allocation, and interaction between humans and machines? How much of the process can or *should* be predetermined and how much should be left open to the interpretation of the actors involved? What roles could these actors assume in an open and interactive construction process? And finally, what would become of digital fabrication if we moved away from predefined optimization problems and towards open design and fabrication processes? These questions informed the trajectory of the research's future development.

Collaborative Design and Construction

My subsequent research explored modes of interactive design and fabrication, such as improvisation. My intention was to move beyond the singular goal of optimization and allow other unanticipated factors to enter into digital fabrication, including human intuition. I strived to innovate a design and construction workflow that evolved naturally using a process of trial and error and continuous reflection.

Improvisation for Design and Construction

The following research was developed under my guidance by Isla Xi Han, a PhD candidate at Princeton University, School of Architecture. Han's research goal was to explore collaborative design and fabrication processes between humans and robots and to attempt to avoid "freezing" a design before it is fabricated. More broadly, Han's project presented an opportunity to explore how openness in human-robot collaboration can lead to unexpected creative outputs.

In contrast to typical construction robotics projects, the proposed fabrication process in Han's work did not follow individual predefined steps; instead, every decision was made based on the results of previous decisions. Of course, it isn't possible to leave all

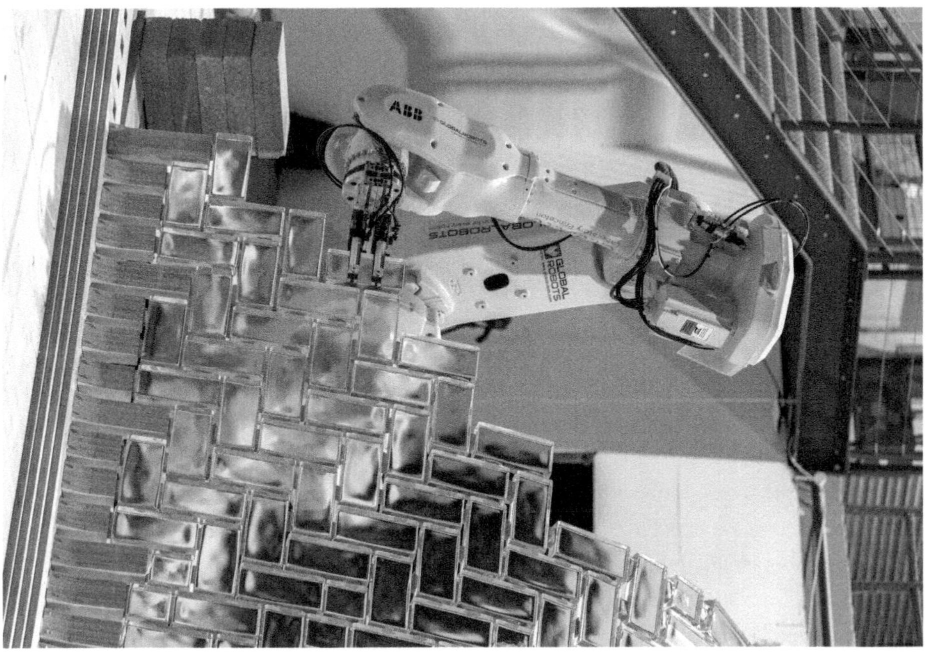

Figure 5.4 Adjustments to the global design required by shortened production time and resulting edge brick conditions. *LightVault*, CREATE Lab Princeton and SOM, 2020.

decision-making open-ended and put the human and the robot in a construction setting without assigning any tasks or roles to either of them. Still, we strove to strike a fine balance between how much of the process we predetermined and how much flexibility we granted each actor. To approach the question of open-ended collaboration more systematically, we defined the individual tasks for humans and machines, but chose not to predetermine the output or limit the roles of both humans and machines to executing solely pre-set tasks.

Figure 5.5 Human-robot collaboration in brick assembly. CREATE Lab Princeton, 2021.

In the first series of experiments, we used robotic block-stacking as a means to explore the impact of robotic constraints on human creativity (Figure 5.5). When stacking different-shaped blocks, the robot—or rather, the algorithm guiding the robot—decided on where to place a new block based on the present state of the structure at any given moment. We started the process by instructing the robot to place a first element close to the bottom edge of a vertical rectangular frame that we had positioned in front of it. Once the robot decided where to place the block, the human collaborator was asked to experiment and find solutions for supporting the robotically placed block by positioning new blocks between it and the existing structure, or the ground. This alternation between human and robot stacking revealed how much the decisions of one actor influenced the decisions of the other. For example, if the robot placed a new block too close to the existing structure, the space for the human-placed block, and with it the space for creativity, was either limited or eliminated entirely. However, if the robot placed a new block too far away from the already stacked structure, the human would struggle to find a solution that provided sufficient structural support. The most interesting discovery for me was that we could influence the predictability or regularity of human's moves by programming the robot's moves to be irregular. For example, if we explicitly asked the robot to place a regularly shaped block in an irregular orientation, the human collaborator would invariably respond with more complex block placements (Figure 5.6).

It is important to note that not every step in the process was successful and that part of the experiment was to explore how the human collaborator reacts to a robotic placement that does not move towards the block-stacking objective. However, since the goal was not framed under the narrative of optimization, we did not evaluate the results by material savings or efficiency in the robotic movement. Instead, we were interested in identifying new creative solutions that might not be *optimal* but might open up a new

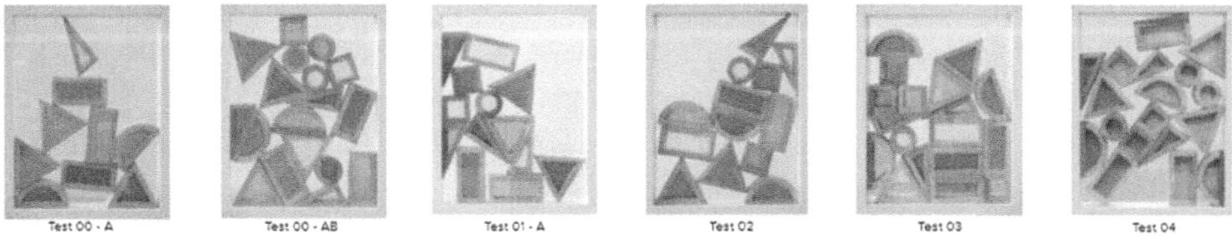

Figure 5.6 Results of prioritizing different placement parameters, such as height, orientation, or colour. CREATE Lab Princeton, 2021.

perspective on a given design problem. In this particular case, we evaluated the results based on the extent of unexpected geometric choices and irregularities that enabled creative outputs that neither the human nor the robot would have achieved alone.

Our experiments in improvisation reveal that there can be more to relationships between humans and machines than a simple division of labour. By allowing the decision-making power to continuously alternate between humans and machines (or algorithm) and leaving the formal design open, we can unlock new solutions to architectural design and construction problems and allow each actor to perceive every new step in the fabrication process as a fresh start. Our intention in this project was not to analyse whether the material results are "better" in one way or another than those of typical robotic construction processes. Instead, we aimed to demonstrate that moving away from purely quantitative goals might open up new ways of creating and evaluating human-robot construction process, which include integrating non-optimal decision-making and acceptance of trial and error.

Conclusion

In the projects presented here, I intended to explore design possibilities beyond those afforded by predefined tasks and instructions. And yet, in these projects, I have admittedly still limited myself to a finite range of pre-set processes and activities. These examples are thus only the beginning of a larger exploration into the balance between defining fabrication tasks and leaving the fabrication processes open to adjustments, adaptation, and intervention. I do not ultimately intend to fully open human-robotic fabrication processes to endless creative possibilities nor do I believe that would be productive. Instead, I endeavour to incrementally *loosen* various fabrication parameters and explore the impact that such technical loosening has on the outputs of both design and construction. In the examples that I shared, the roles of humans and machines are predefined, including the tasks they are allowed to perform. However, we also allowed them to interact, dialogue, build upon each other's work, and even question one another by consciously avoiding a time-dependent end of the design process.

There are two factors that I believe we should consider in order to open up digital fabrication: time and action. If we delay or defer decision-making processes, we can invite more space for adaptability into fabrication processes. For instance, we can extend the time in which we can learn, listen, expand, and adjust; we can, in short, create a more inclusive and lenient fabrication process. Without the pressure of fixing a design goal in a certain amount of time, the fabrication process remains fluid and experimental and makes room for reflection. Ultimately, I'd argue that the biggest danger of optimization-centric design and fabrication research is the inability to question the optimization goal once it has been set. Such inflexibility leads to unproductively clear delineations between failure and success.

If instead we dissolve the roles of the different actors involved in the robotic fabrication process, we can possibly uncover more creative, unexpected outcomes that bear greater potential for reactive and insightful design solutions. Processes that move away from sheer optimization, or at least optimization without context, towards design and fabrication collaboration can revive our currently stagnating field of research. Collaborative fabrication processes can reveal opportunities for designers to use digital technologies in ways that renew fundamental design considerations, such as human interaction, contextuality, and improvisation.

Notes

1 Alison B. Powell, *Undoing Optimization: Civic Action in Smart Cities* (New Haven, CT: Yale University Press, 2021).
2 Sougwen Chung, 'Sougwen Chung: Why I Draw with Robots | TED Talk.' https://www.ted.com/talks/sougwen_chung_why_i_draw_with_robots [accessed 2 October 2022].
3 Madeline Gannon, 'The Robot Whisperer Who Tames Giant Industrial Machine "monsters" to Do Her Bidding', *Wired UK*, 2017. https://www.wired.co.uk/article/madeline-gannon-robotics [accessed 1 July 2022].
4 Stefana Parascho, Isla Xi Han, Alessandro Beghini, Masaaki Miki, Samantha Walker, Edvard Bruun and Sigrid Adriaenssens, 'LightVault: A Design and Robotic Fabrication Method for Complex Masonry Structures', AAG 2020: *Advances in Architectural Geometry, 2021.*

Bibliography

Alison B. Powell. *Undoing Optimization: Civic Action in Smart Cities.* New Haven, CT: Yale University Press, 2021.

Madeline Gannon. 'The Robot Whisperer Who Tames Giant Industrial Machine "monsters" to Do Her Bidding', *Wired UK*, 2017. https://www.wired.co.uk/article/madeline-gannon-robotics [accessed 1 July 2022].

Madeline Gannon. 'The Robot Whisperer Who Tames Giant Industrial Machine "monsters" to Do Her Sougwen Chung, 'Sougwen Chung: Why I Draw with Robots | TED Talk.' https://www.ted.com/talks/sougwen_chung_why_i_draw_with_robots [accessed 2 October 2022].

Stefana Parascho, Isla Xi Han, Alessandro Beghini, Masaaki Miki, Samantha Walker, Edvard Bruun, and Sigrid Adriaenssens. 'LightVault: A Design and Robotic Fabrication Method for Complex Masonry Structures', AAG 2020: *Advances in Architectural Geometry, 2021.*

Wind-Up Architecture

An Interview with Marshall Prado

ZACH COHEN: I'd like to begin with a few questions on interdisciplinarity. You're trained as an architect, but your digital fabrication work seems to draw from several disciplines: architecture, computational design, structural engineering, biology, industrial manufacturing, and probably others. I'd be curious to know how you see your work in relation to these disciplines. For instance, do you feel that your work more closely aligns with one more than another? Or, perhaps, could it be that you're even in the midst of inventing a new, hybrid discipline of your own?!

MARSHALL PRADO: I don't know if it's a new discipline! I think it's more of a blending of disciplines, and that's something that you already see happen quite a bit in architectural practice. No building is just built by an architect: you need engineers, fabricators, and various other trades to come together to make a building. So, yes, I'm trained as an architect, and that's really the lens through which I look at digital fabrication, but the work is very much dependent on multiple disciplines coming together. This coming-together necessitates new ways of communicating and researching, as well as new types of technology; I think my work aims for this kind of invention. At the beginning of a project, I try to consider the expertise that is out there and available, and how it can blend with architecture in a way that builds new architectural opportunities, directions, or constraints. I like to extract the knowledge of different experts—engineers, biologists, computer scientists, etc.—and merge it with my own in order to build a more nuanced understanding of a project than one person alone can have.

DOI: 10.4324/9781003296522-7

Figure 6.1 The UTK Filament Tower at Exhibit Columbus in 2019. © Hadley Fruits.

I also think it's important to clarify that none of the projects that I do exist in the isolated world of architecture; they can provoke conversations on architecture, but they can also provoke conversations on engineering or biology, for example. And, so, the work becomes a little less esoteric because I don't just engage with my colleagues, but also with people that might not even have the same interest in, or understanding of, built space. For instance, if you look at the tower project that I did a couple of years ago as part of the Exhibit Columbus Program in Columbus, Indiana, there were a lot of people that experienced the project that had no expertise in architecture, but they could still come and engage with the project in different ways. I think the work becomes most interesting when it is approachable and instigates more varied conversations (Figure 6.1).

ZC: Absolutely. Let's discuss that tower project: what, in your mind, made that project more approachable apart from demonstrating a blend of different disciplines?

MP: I think part of it was that it was exhibited in a small town in Middle America, and it was really the first time that robotic fabrication technologies had really been on display in that sort of context. It also helped that it wasn't an academic context, but rather a public exhibition, so there was greater exposure to people who had never been exposed to such work before. But I think, even for those who had previous exposure to this kind of work, there were a lot of unique aspects to this project, for example: it was one of the

Figure 6.2 Eliel Saarinen's North Christian Church serves as a conceptual and physical backdrop to the Tower. © Hadley Fruits.

tallest composite projects that has ever been done, it was extremely lightweight, and it was highly materially efficient. I think there were a lot of ways into discussing the work.

ZC: I'd like to ask to specifically about the decision to site the tower next to [Eliel] Saarinen's [North Christian] Church. I found this quote from the Reverend of the Church, who said, in regards to that building, and I'm paraphrasing, that no other piece of architecture could have so well embodied the grandeur and simplicity of the Christian faith. I'm curious to know if you were attempting to make a statement by positioning your twisted tower next to what others have perceived as a pure symbol of Christian faith? Did you intend to commit a kind of "architectural sacrilege"? Or were you, perhaps, offering another kind of symbol or set of values for people to latch on to (Figure 6.2)?

MP: I think the tower typology certainly embodies, or stands for, something. When you go to Columbus, you see these iconic towers, like the Saarinen church or the Venturi Scott Brown fire station. I was definitely looking at that architectural history present within the town and trying to think how I could engage with that culture.

I actually hadn't heard that quote from the Reverend, pretty interesting. I'd love to have a conversation about what the tower meant to the congregation or what he thought about it. I do think I had ambitions of it being a beacon for the area.

I also think that there was generally this dichotomy of daytime and nighttime, or dark and light, that we designed into the project, even down to its materiality. The carbon fiber, for example, is opaque and strong, and the glass fiber is like an optic fiber: it glows and highlights its silhouette. This material interplay created this back and forth (or even good and evil!) sort of situation, but it plays on visual cues that can be read in different ways. So whether it's architectural or religious or a structure purely for research, the most important thing to me is to build in various interpretations.

I remember one of the first carbon fiber-winding projects I did at the Institute for Computational Design and Construction (ICD) in Stuttgart, Germany. We were in the middle of installing it and this young German boy came by and started to walk through it and said "Spiderman hat das gemacht," or, "Spiderman made this." It was really interesting to hear him have this kind of connotation, like it was a project made by a superhero, or just otherworldly in some way. I like to hear those kinds of (mis)readings (Figure 6.3).

ZC: It's interesting to hear you describe the tower that way because it does seem to embody a kind of duality: a struggle between this pure white and this pitch black that plays out through the structure.

MP: They certainly play different roles in terms of both structure and graphics and create new readings of the architecture. For example, people often have an expectation of an architectural structure being something massive, like cast concrete. But the tower's lightness challenged that expectation. I think about the people that witnessed the install and saw these gigantic carbon fiber components being easily picked up and assembled. The act of building becomes almost superhuman in the sense that these kinds of robotically fabricated structures can be assembled without the ungainly infrastructure and machinery that people typically think construction requires.

ZC: Do you think that it's important that robotically fabricated structures demonstrate this kind of superhuman, or even posthuman, element in both process and product? For example, what if you used your highly advanced fabrication methods to produce really generic things instead of alien-looking towers? What if you just fabricated something simple like a box? Is it important that the extra-humanness that goes into the making is also expressed through the robotically fabricated thing in some way (Figure 6.4)?

MP: These kinds of tools are most often used for efficiency's sake; for instance, if they are used in a non-architectural setting, or in a production capacity, people are mostly using them to fabricate boxes, or other simple things, efficiently. This kind of approach is a holdover of the standardization and mass production that has dominated manufacturing for hundreds of years. But for me, that's a missed opportunity. I think we need to ask ourselves, what sort of material opportunities are afforded by these emergent tools? Where can these new technologies lead us? And I think that's where the expressive part of the work comes in.

I would also like to think that even if I had no technology, I would still be exploring the potential of new forms, materials, and processes; that's fundamental

Figure 6.3 The Tower glows at night to provide a beacon for the surrounding town. © Hadley Fruits.

to my work. And even though I do use "high-tech" machines, I try to simplify their application as much as possible. For instance, I try to use off-the-shelf parts wherever possible. I try not to integrate the technology where it's not needed.

ZC: Would you be able you give us an example of an architectural application of robotic fabrication where you thought to yourself, "I don't think the technology is needed there"?

MP: That's a good question. Especially within academic research, there's a constant fight to innovate and discover ways that technology can be applied to existing processes rather than attempting to discover ways in which technology can simplify those

Figure 6.4 The "heroic" carbon fiber construction of the 2013 ICD/ITKE Research Pavilion at the University of Stuttgart. © ICD/ITKE University of Stuttgart.

processes. So, in the end, the desire for innovation sometimes leads to an overcomplication of ideas. That doesn't necessarily mean you shouldn't attempt to incorporate technologies, because you wouldn't otherwise be able to know if and when they're needed. But I do often find that if I try to use technologies in ways that they're not needed, there needs to be an evaluation process to understand that. I can't point to projects that other people have done because I wasn't involved in this kind of evaluation, but let's take one of the projects I worked on at the ICD. I should also preface this by saying I don't mean to discredit the work of my collaborators in this project because they're constantly pushing the bounds of the discipline and should be congratulated for all the risks they take.

In the 2016–2017 ICD/ITKE Research Pavilion, we aimed to use a team comprising a robotic arm and aerial drones to fabricate a cantilevering carbon fiber structure. One goal of the project was to expand the workspace of a single robotic arm, which is generally very limited. The addition of the drones gave the robotic arm the potential to "reach" much further away and therefore design a much larger structure. However, the actual implementation of the drone technology—a technology that can range over square miles with ease—on a relatively small demonstrator was almost entirely unnecessary. We could have easily made the same demonstrator without using that technology. But the demonstrator (the pavilion) was not the output of the project. The output was to demonstrate this robotic arm-drone system. And this system does have the potential to be scaled up; for example, if you wanted to use this system to wrap a stadium roof or similar long span structure. But at this scale—the scale of the pavilion—the technology was actually more of a hindrance than a help.

In the end, whether or not the technology is useful or appropriate depends on how you evaluate the project. For instance, one thing we discovered from the cantilevering pavilion project was that this robotic arm-drone collaboration was actually a very slow process, and that the process could be sped up if we did some of the fabrication work manually or "robotically assisted." In another project, of a different scale, this human-robot collaborative labor might be something to consider. But when we were developing this project, we had ambitions of industrialization and transferring this system to larger-scale applications, and I think that ambition ended up framing the project in a different way—one in which using drones might be an entirely suitable way of working. If you're making a large-scale tower, for example, like the one I made in Columbus, or if you're working at the scale of a skyscraper, different technologies might need to be developed and embedded in the system that can increase scale. I think if we're going to seriously evaluate the ways in which robotic fabrication technologies can change the building industry then we need to evaluate them at the scale of a building (Figure 6.5).

ZC: You mentioned the necessity of establishing evaluative criteria and the ways in which the success of the research inevitably varies depending on the evaluative criteria being used. Does that criteria change for you from project to project, or is there a set of criteria that you consistently return to?

Figure 6.5 The robotic arm-drone fabrication system demonstrated by the 2017 ICD/ITKE Research Pavilion at the University of Stuttgart. © ICD/ITKE University of Stuttgart.

MP: All of my projects are under the umbrella of one sort of meta-project, which involves composites, robotic fabrication, computational design, and lightweight structures. For me, each of these components is a different way to address the same overarching project. For example, when you get into questions of geometric freedom or structural optimization or integrating building systems, each facet of the overarching, meta-project will speak to these questions differently. The facet that does the speaking, let's say, changes between individual projects: a component that we might not have thought about at all in one project can become the focus of another project.

You can see that kind of nonlinear conceptual development in the very first ICD/ITKE Research Pavilion in 2013. We were looking at developing a system to create highly geometrically differentiated components without a mold. In initially demonstrating this system, it was pretty important to show its potential to disrupt serialized production. Then we used the same system at the Victoria Albert Museum in 2016 and created the Elytra Filament Pavilion. In that pavilion, the geometry of the individual components was relatively similar, at least in profile, but we were exploring variations in the winding sequence to give each component different structural performance, apertures, or functional integration. And so, that criterion, which was nonexistent in the initial pavilion, became the focus and, as a result, shifted the project quite a bit.

I think that it's a constantly moving needle. If I start a project and feel that I'm not addressing new things, then there might be no reason to do it (Figure 6.6)!

ZC: Let's pivot to discussing winding more specifically because that's clearly a fabrication method that you've evolved through several of your projects, like the Elytra Pavilion. Can you speak about the origins of winding in your work, specifically your interest in robotically-wound carbon fiber?

MP: To begin with, it's worth saying that the technique is not new: filament winding has been around for quite a while, but the process is usually limited to producing very specific geometries or using a mold. The reason why filament winding is interesting for architecture is that it's one of the most efficient forms of composite production: it's fast and grants fabricators a lot of control over the material itself. It's also very different for architects to think about materials from a very small scale in which they can actually build up the material performance as they go, and that's what they're able to do in filament winding processes.

For instance, when you talk about using 2×4's or sheets of plywood, the material embedded in those components already comes prearranged, and so, you need to use a subtractive process to put these things together. In contrast, I can use filament winding to build up material performance, determine grain directionality, or control how each component will react under certain loading conditions. When you're laying material fiber by fiber, you can establish different performances simply by modifying how those fibers intersect—these are the kind of material principles that we see in nature. You can look at the microscopic material makeup of trees or bones and you will find a dense network of fibers. We can't grow our own architecture in the same sense that natural organisms grow, but we can understand those ideas of material arrangement and design processes that enable a similar kind of growth.

Figure 6.6 The Elytra Pavilion at the Victoria Albert Museum was constructed through a variable carbon fiber winding process that embedded different architectural performances into the composite components. © ICD/ITKE University of Stuttgart.

Figure 6.7 The carbon fiber winding process, and its resultant materiality, can be both robotically and manually tuned. © ICD/ITKE University of Stuttgart.

ZC: Then would it be fair to say that one of your main interests in the winding process is that it allows a finer tuning of material performance (Figure 6.7)?

MP: Yes, fine tuning. When you're not just assembling objects, but you're winding materials, you can fine tune across scales.

ZC: Can you speak about what role historical notions of craft play in your research? I ask because you're taking these thousand-year old processes, like winding, that are quintessential examples of human craft and are automating them—replacing humans hands with robotic arms. Do you think about what is gained and what is lost in the process of automating such crafts? Or is it more, like we spoke about earlier, that you're thinking about the affordances of these technologies and you want to see what forms of winding they can produce, good or bad?

MP: I think each process we develop starts analog before it moves into an automated or robotic process. There is some sort of initial testing to understand how the analog process works and if it's even suitable for automation. For example, most of our winding systems start with a model: we string fibers to a frame and think, could this be applicable to a robotic process? And what are the benefits of it becoming a robotic process? In the end, there's always a negotiation between the analog and the automated. We're always trying to understand the process of winding and then think about how automation could create a variation of that process; not necessarily to automate for automation's sake, but as a way of scaling certain crafts up to architectural processes. The way to advance any craft is to understand what the tools you're using give you. Automation gives us another way to understand tools and tool affordances.

Now, do you lose some of the craft with automation? The individuality of human paintbrush strokes, let's say, cannot be expressed in an automated way with as much embedded personality. That might be something to think about in terms of craft-based labor. But you gain other things, especially when you apply automation in a different labor context like architecture.

I often get asked why am I using a robot if I can wind these things myself. And the answer is, it never turns out as good as when a robot does it! There are certain things robots can do better than people in terms of speed, strength, and consistency.

ZC: I think this idea of automation as a tool to evolve craft is interesting. If I understand what you're saying correctly, it sounds like attempting to automate certain fabrication processes, even if it's not productive from an industrial standpoint, can be productive from an epistemological standpoint; in other words, automating crafts can simply enhance our knowledge of those crafts. Would that be fair to say? If so, it then makes me wonder if there is, in fact, something like "robotic craft" or if robotic fabrication is mainly for the sake of furthering our own understanding of *human* craft (Figure 6.8)?

MP: I think it can, but I also think there *is* a robotic craft; it really depends on how robotic fabrication is approached and whether or not someone is approaching it in terms of craft and material understanding. The machines themselves are used to do things that are not about craft but maybe about consistency or speed. Those concepts in themselves do not readily conjure notions of craft, but that doesn't mean that we,

Figure 6.8 The carbon fiber winding process enables a synthesis of human and robotic craft that puts forth a new kind of optimized architecture. © NAARO.

as designers, can't bring craft to the table. For instance, you can use these tools in ways that are about optimization of certain production methods without necessarily exploring their potential to enhance different material processes. I think when this latter kind of exploration enters into the work, that's when a robotic craft starts to emerge.

ZC: Thinking about the possibility of a robotic craft also makes me think about the possibility of scaling up, which you've repeatedly referred to as a driving force behind your research. Scaling up and developing a robotic craft seem to be contrasting pursuits. When we think of craft, I think we tend to think of a certain, perhaps more human, scale. And it seems, in fact, that in digital fabrication's 20- to 30-year history in architecture, it has been more or less stuck at this craft-oriented scale; one only needs to look at the dozens of robotically fabricated pavilions to see that this is the case. To me, that suggests that we've so far been more interested in developing new crafts instead of reaching larger scales. What do you think it's going to take for architecture-oriented digital fabrication to *actually* scale up? And secondly, what scale do you think it can reach?

MP: I'm not sure the main hurdle is scale as much as it is application in the field. I think the issue with pavilions, for instance, is that they inherently, even deliberately, limit many of the constraints an architect may encounter in practice: for example, there

Figure 6.9 The Elytra Pavilion presents possibilities for scaling up, both in its modular construction and its greater engagement with the bureaucratic and physical inherent to most non-robotic building practices. © NAARO.

are no clients, builders, or programs. When those kinds of digitally-fabricated pavilions have tried to contend with these additional project inputs, their innovative or interesting aspects often get subdued—what begins as a simple exercise in fabrication becomes a much dirtier problem.

Something you also see a lot is when really innovative digital fabrication research gets spun into a startup, it has to confront things like building codes or bureaucratic processes, and it immediately becomes limited in what it can do. These sort of obstacles are exactly what have stagnated change and innovation across the building industry (Figure 6.9).

ZC: Would you say that digital fabrication is, in a way, destined to lose its "superpowers" once it exits the realm of academia? In other words, when it's being incubated in academia, digital fabrication has all this rich potential that we love to explore, but as soon as it leaves that incubator and encounters the fictious elements of the construction industry, it seems to immediately become diluted. It makes me wonder if digital fabrication's greatest benefit to architects is that it presents us with new potentials. And even if it doesn't fulfill those potentials, maybe that's okay. The potentials themselves could be enough?

MP: I think there could be a future in which it becomes more applicable. I don't think it's a matter of architectural academia needing to change as much as the goals of the building industry needing to change. For example, look at composites in other industries, like automotive or aerospace engineering: they've become the go-to material because they provide a material efficiency and performance that aren't available in other materials. In contrast, one of the biggest issues with the building industry is that it's extremely wasteful in terms of its material usage. So until material performance or material resourcefulness become key concerns of builders, composites probably won't become a heavily applied system. When material supplies start to run out, and the question of material reuse becomes foregrounded in building production, then I think composite fabrication will be ripe for application. I don't think the consistent production of pavilions will outlive the novelty of the material itself (Figure 6.10).

Figure 6.10 The spectacular oculi of the UTK Filament Tower demonstrates the ways in which composites can be both efficient and performative. © Hadley Fruits.

Dissimulated Origins

On Navels and Nozzles

Francesca Hughes

We are wont to forget that Vitruvius never actually drew the Vitruvian man; or at least, if he did, no drawing has survived. Hence, ironically, the location of the point that pinned down Western architecture's home coordinates and the generative geometric origin of all virtual and material line-making that ensued is, as translator of Filarete, John R. Spencer, describes, "strict but rather vague."[1] *Strict* in that if one follows Vitruvius' instruction to the letter man fits perfectly in a square and in a circle. *Vague* in that the centre of man in the circle is his navel, while the centre of man in the square is, as Joseph Rykwert discretely puts it, his "generative member."[2] C. Cesarino's 1521 woodcuts of Vitruvius' bi-centred men present both origin points as equally valid, apparently without conflict, suggesting navels and genitals are interchangeable. And they are. Architecture's autopoietic home conceals (barely) the unstable relations to origin coordinates that anchor the sexual metaphoric acrobatics at play in its reproductive logics[3] (Figures 7.1 and 7.2).

Even if we overlook this inconsistency and stay with the navel as centre of the circle, this too is constitutionally unmoored. In a quiet aside in Filarete's *Treatise on Architecture* he admits to the possible discovery (upon testing with his compass) that the origin or centre point is adrift. In asserting all measure is derived from the human body and duly reciting Vitruvius, he adds:

> Vitruvius says that the navel is the middle of the figure of man. The navel would be the center for the point of a compass that would circumscribe the man in a circle. From this the circle derives. It is a good enough reason to

DOI: 10.4324/9781003296522-8

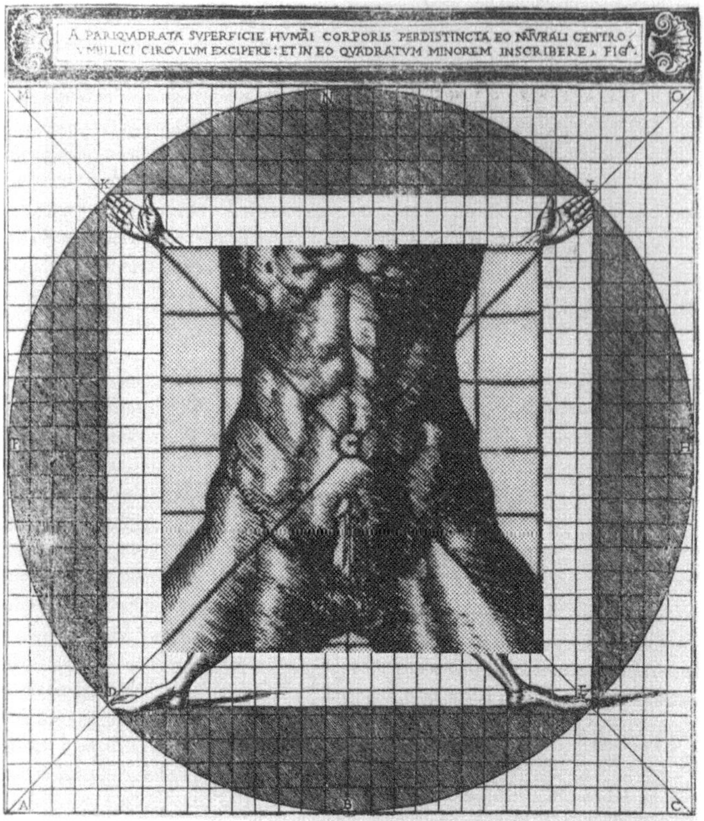

Figure 7.1 Author's composite: Woodcut of man in circle, Cesare di Lorenzo's (Cesariano's) edition of Vitruvius' *de Architectura* (1521).

> confirm our proposition that all measures are derived from man. However, *it does not seem to be exactly in the middle*, but the circle does *remain round*.[4]

How can this be? What is a circle that is able to 'remain round' while its compass point is not 'exactly in the middle'? If the two locations that determine the generative home of the building, the navel as origin point and the centre of the circle the same navel's body describes, do not align, are we to assume some silent arrangement has been reached?

The homing of the machine or robot, at whose tool-centre-point (be it via addition or subtraction or a combination of the two) a building now 'arrives', inevitably extrudes this legacy of dissimulated ambiguity.[5] Vitruvius' navel is the origin point in space (and time) from which architecture begins its making, the coordinates that serve as reference point, as (0,0,0), for all subsequent coordinates in the drawing and then the materialisation of building. The navel is the home position to which architectural bodies must serially rehome so that neither line nor labour nor material gets lost along their 'tool-path'. Just as homing (or referencing) allows the machine to internally figure out where the end-effector or nozzle-head is located in machine-space, the projected and projecting navel in the basilica's section, the church's plan and the city's organisation,

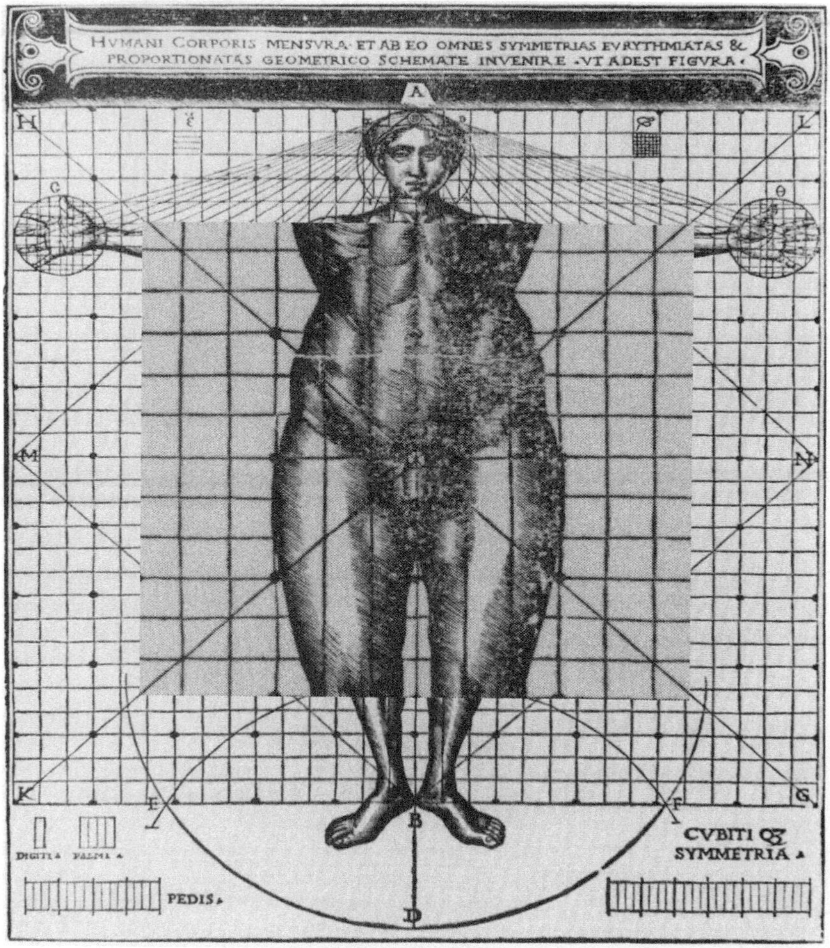

Figure 7.2 Author's composite: Woodcut of man in square, Cesare di Lorenzo's (Cesariano's) edition of Vitruvius' *de Architectura* (1521).

allowed roof tile, altar slab and lawcourt to figure out where they were located with respect to a global—but ultimately hermetic—geometric logic whose (0,0) anchored all. Not only does the latter-day Vitruvian navel also determine all, it comes first: the first thing a CNC router must do before departing down any milling path, is home itself so that it does not err along the way. Like all world-making, the machine is drawing its world from its own point of view.[6] As Thomas Nagel reminds us however, the view from the machine is the view from nowhere.[7] So what does it mean to get lost within a system that is itself already organised from (and by) nowhere—is already lost? What if we were to allow the oozing nozzle to get led astray in its material wanderings, to err as it senses, in the closeness of the closedness, and submit to the feedback from the material it is depositing? What navigational exchanges would then be negotiated between (the frictional whims of) formlessness and the view from nowhere?

Returning by way of Spencer's "strict but rather vague" logics that describe the meeting of the corporeal with the geometric in production, we find a latter-day echo

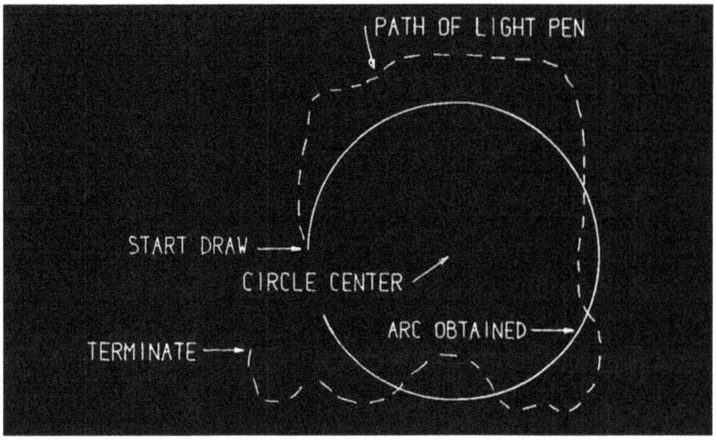

Figure 7.3 Sketch by Ivan Sutherland explaining how to draw a circle in *Sketchpad* with the light-pen showing how the program's constraint logic ignores the erratic radius, accepting only the radial dimension of the first point on the circumference and subsequently inputting only the degree of arc drawn. Sutherland, *Sketchpad, a Man-machine Graphical Communication System* (Cambridge: MIT, 1963, p14).

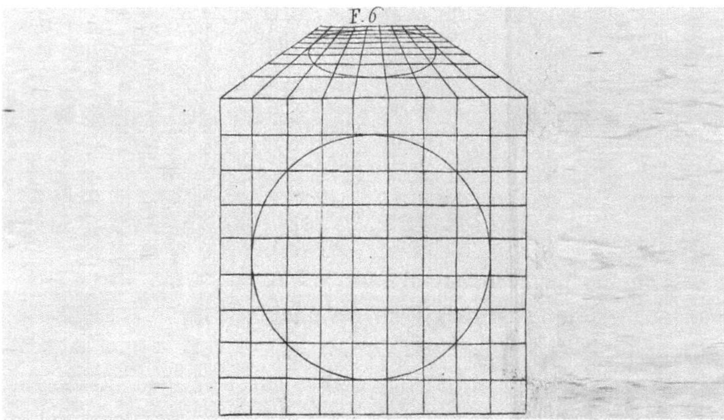

Figure 7.4 Figure from Alberti's 1435 *de Pictura* explaining how to erect a scaffold to draw a circle in perspectival projection. Leon Battista Alberti, *Della pittura e della statua di Leonbatista Alberti* (Milan: Società tipografica de'Classici italiani, 1804, p 179).

of this paradoxical construction in Ivan Sutherland's 1963 design of an interface that allows a hand to draw a perfect circle (Figures 7.3 and 7.4).[8] Under his algorithm, the gesturing sweep is allowed to be "quite sloppy and still get a precision drawing out at the same time."[9] Sutherland's update to Leon Battista Alberti's circle-drawing system, issued some 500 years earlier in *De Pictura*, completed the decorporalisation of circle-drawing by erasing the last trace of the body altogether. Whereas Alberti had devised a way to draw radial perfection by erecting an approximating join-the-dots scaffold, Sutherland did away with the stepping-stone dots altogether.[10] Alberti's armature gently corrected the erratic radius of our hand; *Sketchpad: a Man-machine Graphical Communication System* effectively ignored it and used the input tool (the light-pen's) location *only* to decide how much arc to draw:

> To draw a circle we place the light pen where the center is to be and press the button 'circle center,' leaving behind a center point. Now, choosing a point on the circle (which fixes the radius) we press the button 'draw' again, this time getting a circle arc *whose angular length only* is controlled by light-pen position.[11]

Angular length not radius: in order to eliminate the body, the interface design is such that the body drawing a circle in *Sketchpad* does not have to (be able to) draw a circle to draw a circle.

Returning by way of the covert ambiguity of the circle that is able to 'remain round' while its compass point is not 'exactly in the middle' and the overt ambiguity of our bi-centred men, the 'home' position is usually not 'machine zero'. In the case of a robot the local centre is likely the 'armpit' or even 'elbow' condition of the fabricating 'arm'; each a contextualised offset from the (0,0) of the robot itself that echoes the relations between the 'generative member' (an adaptation for the square) and the navel—if we accept the circle as the most fundamental geometry. The 'hand' (wrist), or tool-centre-point, becomes an even more local centre in reverse kinematics—how to tell the drill bit to be perpendicular to the surface it is drilling for example. But what if the 'hand' acquires Copernican ambitions, fixes itself and instead requires the 'navel', the entire robot, to rotate around it? Where is nowhere now?

Returning by way of the body anchored by its navel and its respective umbilical logics, the co-nascency of two 1960s nowheres is important here (Figure 7.5). Two years after the first objects drawn in Ivan Sutherland's *Sketchpad* were being 'rotated', 'tipped' and 'shifted' at MIT's Lincoln Laboratory, astronaut Ed White used 'pitch', 'yaw' and 'translation' to narrate the relocation in space of his umbilically defined body in the first ever EVA (extravehicular activity) or 'space-walk', while floating at the end of an oxygen tube whose other end was attached to the equally unmoored Gemini 4. He later narrated his own filmed transformation matrix, "I thought in my own mind that I could control myself in both pitch, yaw and translation. This is the type of control that you need to move from point A to point B in space."[12] White's translation also has to get him home:

> Relay down that control with the gun was very good in yaw, very good in pitch. Roll control wasn't good but when I had the gun, I was able to pick a spot on the spacecraft and return to it. I was able to maneuver [sic] in translation.[13]

What, one might ask, would Florentine Filarete and his neighbouring Sienese Francesco di Giorgio, with their umbilical fixations, have made of the decidedly adrift White, and the meandering nozzles from which a building now oozes? With their preoccupation with

Figure 7.5 Author's composite: Astronaut Edward White in the first American EVA June 3, 1965. 19:54 UT (NASA photo ID S65-30433) & Francesco di Giorgio, city as man, Folio 1, Codex Saluzzioanus, Turin.

formal and material points of origin, they may well have recognised several salient features. Even more so in Di Giorgio's case for whom (in the *Magliabecchiano* manuscript) the Vitruvian navel in the design of a city is rendered all the more complexly generative as no longer simply the geometric and organisational origin point but also the point where inorganic material and biomass (like oxygen to the astronaut) enter to feed the system:

> I therefore say first that the major square [piazza] should be in the middle and center of the city (or as near as possible), located as is the umbilicus in the human body.... and the reason for this analogy is the following: since human nature first takes all its nourishment and perfection through the umbilicus, so through the [principle] public square all the other places [of the city] are nourished.

In its newly enlarged scope—'nourishment' and 'perfection', biomass and geometry—the navel as centre holds. He continues,

> and there is justification from nature: that all things which are in common should be equally available to all, as the center [is equidistant] from every part of the circumference. It should moreover be surrounded by stores and honourable businesses.[14]

Within Francisco di Giorgio's schema, all power is to the navel as geometric, reproductive and infrastructural (nutritional) home. This home is also architecture's latest rehearsal of male metaphoric appropriation of female reproductive capacities.[15] As Diana Agrest points out, Di Giorgio's male navel as generative point is recast as the female womb—or, as supply of material, as proto-placenta to be precise.[16] This is a logical elaboration of the metaphoric autopoietic infrastructure Filarete had erected more than two decades earlier in the same treatise as that in which the drifting navel had so perplexed him:

> You perhaps could say, 'you have told me that the building is similar to man. Therefore, if this is so it needs to be conceived and then born'. As [it is] with man himself, so [it is] with the Building. First it is conceived, using a simile such as you can understand, and then it is born. The mother delivers her child at the term of nine months, or sometimes seven; by care and in good order she makes him grow.

Thus the architect is the (male) 'mother', to whom the building's navel is 'umbilically' tethered. We then learn who the father is:

> 'Tell me, how is this conception achieved?'
> 'The building is conceived in this manner. Since no one can conceive himself without a woman, by another simile, the building cannot be conceived by one man alone. As it cannot be done without woman, so he who wishes to build needs an architect. He conceives it with him and then the architect carries it. When the architect has given birth, he becomes the mother of the building. Before the architect gives birth, he should dream about his conception, think about it, and *turn it over* in his mind in many ways for seven to nine months, just as a woman carries her child in her body for seven to nine months'.[17]

Turn it over—long before it has a material life the building is a floating body whose freedom of movement can only be described by a translation matrix. Like the quickening that is becoming, what is at stake in this turning that is making? Representation will ultimately fix the architect's turning conception

> He should also make various drawings of this conception that he has made with the patron, according to his own desires. As the woman can do nothing without the man, so the architect is the mother to carry this conception. When he has pondered and considered and thought [about it] in many ways, he ought to choose [according to his own desires], what seems most suitable and most beautiful to him according to the terms of the patron. When this birth is accomplished, that is when he has made, in wood, a small relief design of its final form, measured and proportioned to the finished building, then he shows it to the father.[18]

The building then is 'born' in the space of representation, a wooden model.

Returning by way of the feeding that delivers nourishment and perfection, Filarete continues: "As I have compared the architect to the mother, he also needs to be a nurse. He is both mother and nurse."[19] The architect/nurse and nascent model/infant make up the closed system of need and satisfaction that defines the oral stage to the point that, as Donald Winnicott concluded, "there is no such thing as a baby, there is only a mother and a baby."[20] In his description of this circuit Mark Cousins used the term *machine* to describe the psychoanalytical apparatus in play:

> One should think of them [mother and infant, here architect and model] as an undifferentiated machine. I don't mean machine in an engineering sense, but what the French call *machine*, an apparatus in which the fundamental experience is that of 'taking in', or not 'taking in'.[21]

A binary logic also underpins this *machine*. Attaining completeness within these relations is ever-precarious: Winnicott would have agreed with Filarete, the architect/nurse need simply be a 'good enough mother',

> As the mother is full of love for her son, so he will rear it with love and diligence, cause it to grow, and bring it to completion if it is possible; if not he will leave it ordered in such a way that it will not perish because of its incompleteness.[22]

Figure 7.6 Author's composite: Zach Cohen, *Dripped Corner*, photograph Sarah Wagner, 2018 & Stuart Brisley, *The Collection of Ordure*, photograph Stuart Brisley, The Freud Museum London, 2002.

Returning by way of the nozzle that now excretes corpulent lines between its pregnant pauses, we are reminded by Cousins that "the anus is perhaps the most important organ in the education of the architect" (Figure 7.6).[23] The nozzle (as vicarious anus) gives its excretions generously:

> It is, as it were, the infant's first great gift to externality. On the one hand the anus discovers the external world. And the first generous impulse of the infant is to give the world the best that it knows to give.[24]

In so depositing, the (closed-loop) robot, as our (projective) yet-to-be-repressed infant, discovers, to its very great delight, the outside, and thus, the inside also: "This pleasure is the discovery of the difference between the inside and the outside."[25]

Crucially, as Cousins argues further, for the infant—as for our robot—"the passing of an object from the inside to the outside has still been done without reference to space," adding, "so we can begin to articulate a second precondition, that the distinction between the inside and the outside is *prior to* the production of space in general."[26] Thus, contrary to the many definitions of architecture that insist upon the idea of inside and outside being predicated upon the prior idea of undifferentiated space and its subsequent enclosure, Cousins' spatial analysis of the anal stage argues that conversely "the differentiation [between inside and outside] is what opens the route to space—rather than the idea that space needed to pre-exist before inside and outside could be imagined."[27] The lesson of the nozzle/anus is clear: the idea of space first emerges from excretion, not enclosure. This time however, the attendant delight is not interrupted by rejection: no parent is repelled by the depositions.[28] On the contrary, while, like the anus, the robot's orifice does again become a site of extreme control, this time it is not organised around the question of disgust—the disgust that *writes* civilisation—but around the pleasure of autoeroticism. Cousins sets out the relations between autoeroticism and extreme control:

> According to Freud, if at any time we cannot gain sufficient pleasure from one stage, we will go back–what psychoanalysis calls regression–to a previous stage. We will regress from love to autoeroticism. If that fails to catch our fantasy, we can relate to objects through extreme control,

adding, "and if all else fails, we can put our thumb in our mouth."[29] However, the excreting robot instead presents an anal phase that does not conclude (if anal phases ever do) with repression and counter-rebellion, but with the eternal pleasure of production. We are embodied anew in the making of the building.[30]

Returning by way of Filarete's other homing coordinate, as materialisation of the building commences, the regime of navel and nutrition becomes the new regime of the generative member and desire as his architect's 'maternal' love quickly becomes erotic:

> Building is nothing more than a voluptuous pleasure, like that of a man in love. Anyone who has experienced it knows that there is so much pleasure and desire in building that however much a man does, he wants to do more. Sometimes he is never concerned with the expense; examples of this are seen every day. When a man is in love, he gladly goes to see his beloved. When she is in a place where he can see her, he is not sorry for the time spent nor is he bored.[31]

Cousins reminds us in his spatial analysis of Freud's stages in the archaeological stratification of sexual maturation that 'sexuality' stands for 'relations to the object'.[32] With the third, phallic, stage arrives the pleasure of objecthood:

> It is the moment at which the subject can take his or her own body as an object. It is the first time that the subject begins to recognise that I can be both subject and object for myself. I can be both the subjective wish for pleasure and I can be the organ object, which delivers this pleasure.[33]

In a curious inversion of homing, Filarete's amorous architect returns and returns to the object of desire, as if tethered to the pleasure they supply:

> So he who builds goes gladly to see his building, and as often as he sees it the more he wants to see it and the more his heart swells. Time passes and he is never reluctant to look at it or to talk about it exactly as a man in love talking about his beloved. There is no half way for him; he loves it.[34]

Returning by way of Sutherland's gesturing hand newly untethered by radius, is there not some corporeal presence in these shapeshifting excretions and their sexual confusion? Has the body, as the previous home of thought quietly reinstalled itself in the corporeal productions of the new machines of thought? Has another arrangement been made? Ramon Llull's 13th century project to cut off the production of truth from the senses, and thus from the corruptions of bodies and their creaturely limitations (i.e., sin), saw decorporealisation as a necessary first step in thought getting ready to be housed in another kind of body, yet to be imagined—the machine of thought (Figures 7.7 and 7.8).[35] Although the algorithm, with its autopoietic fertility, was already there.[36] As too was the permutational paper wheel device Llull had developed (a paper 'computer' of sorts) that could calculate exhaustively and *ad infinitum*, uncontaminated, at last, by the errors that plague human cognition (and flesh). Llull's *Ars Magna* sets out the terms for third-degree knowledge—knowledge utterly divorced from the sensorial—as the production of knowledge decorporealised. This is the promise of thought that will think no falsehoods. If first-degree knowledge was understood as sensible knowledge from sensible things, and Aristotle's second-degree knowledge, in which perception provides input data before abstract reasoning takes over, as intellectual knowledge of sensible things, then Llull's *Logica Nova*, in excluding perception entirely, produced third-degree knowledge—intellectual-only knowledge of intellectual-only things. In so doing Llull completed the disembodying of thought that Aristotle, as conveyed through Averroes, had started. The decisions of his rotating paper wheels were devoid of corporeal input. But, as the wheels already suggest, disembodied thought must still *move* somehow and, in this instance, curiously chose to turn. Not least in Llull's ternary system was its triangulation of feedback: the knower, the known and the act of knowing, or, as Llull also describes it, more erotically as the lover, the beloved and love itself. To move, thought decorporealised must be lent some other body, but a body-without-senses, a senseless body, this time. Just as the relations between inside and outside can be understood as occurring prior to the idea

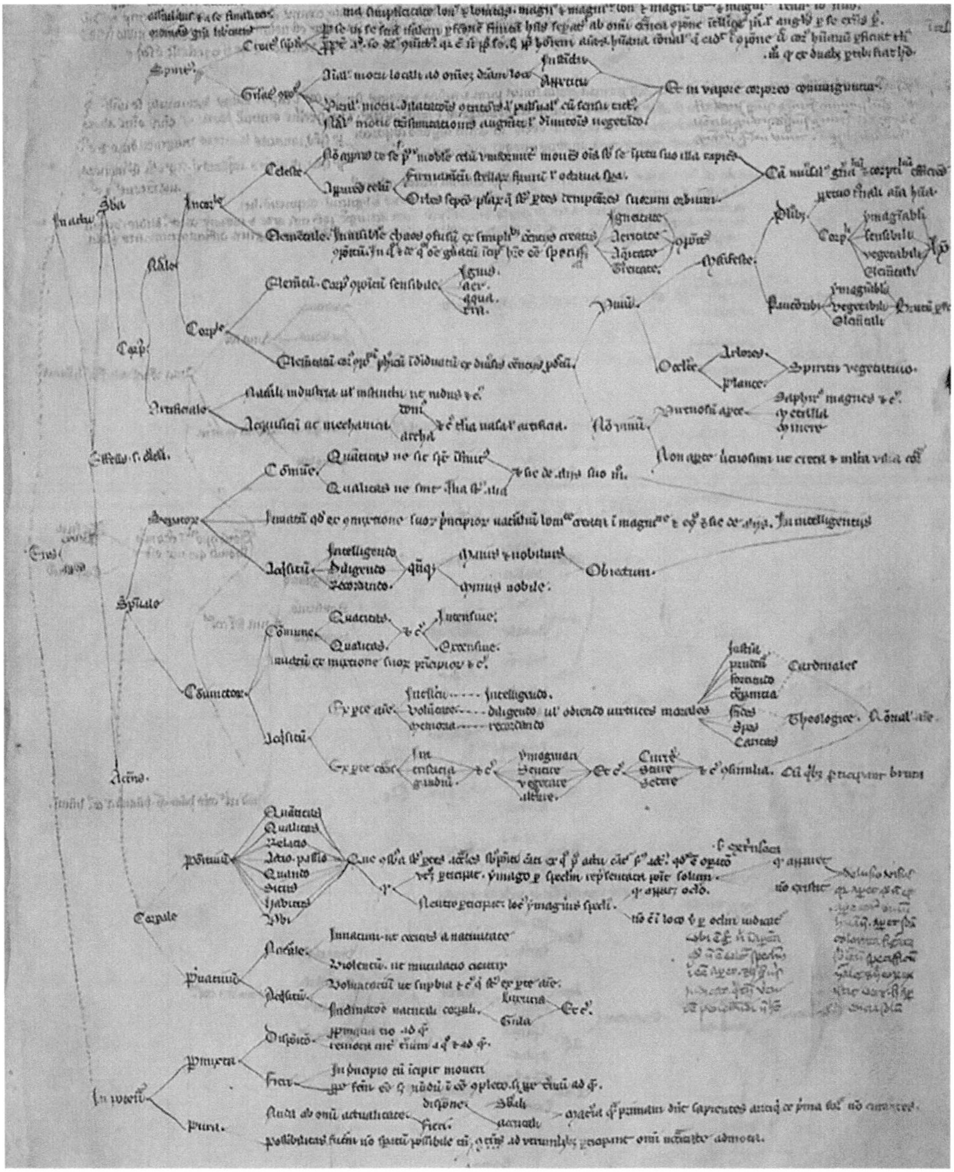

Figure 7.7 Ramon Llull, *Breviculum*, detail, Badische Landesbibliothek, code: St. Peter perg. 92, Karlsruhe.

of undifferentiated space, the decorporealisation of thought can be understood as occurring *prior* to arrival of machine of thought, in which decorporealised thought is subsequently rehoused. That is, thought was not decorporealised in order to install it in the machine: the rehousing of thought in the machine *resulted* from Llull's decoupling of thought from the body—Aristotle's senses—in order to hermetically seal it within the intellect—of the intellect, by the intellect, with the intellect—and yet still satisfy its need to move, to process.

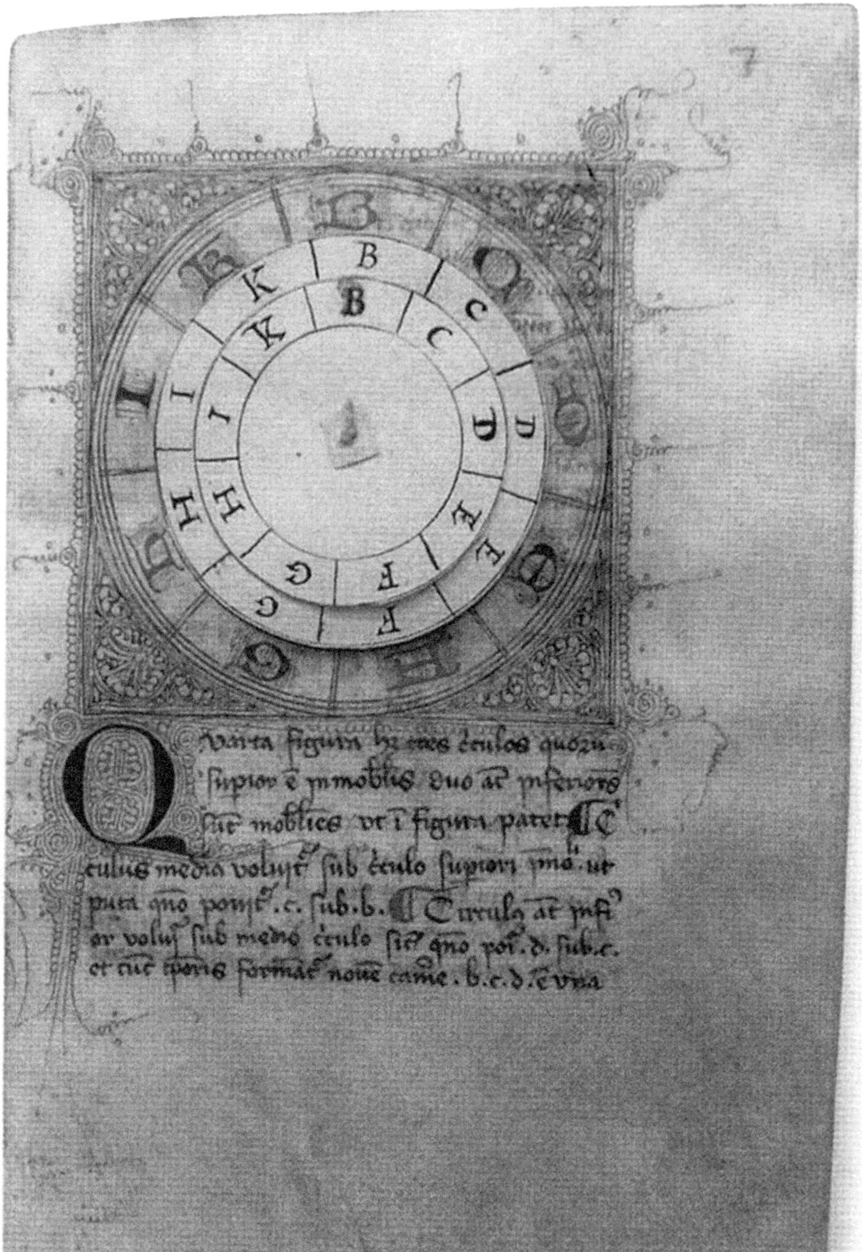

Figure 7.8 Ramon Llull, 'The Fourth Figure', *Ars Brevis*, 1308, Colecciones Reales, Patrimonio Nacional, Biblioteca del Real Monasterio de San Lorenzo de El Escorial. S. XIV.

Returning by way of sexuality as relations to objecthood, a robot crawls through a hollow section of the arch of Sydney Harbour Bridge, repairing as it goes.[37] A portable navel, an RFID tag which it will move as it travels, gives it a home. A cable supplies it with power in return for data. A camera gives it eyes. It recalibrates its position as it scans its environment. The deposition pile or subtractive void it contributes serve to define where its inside meets its outside as it builds an occupancy map in the nothingness. 'Place' is a probability model where greater resolution is greater determinacy. 'Objecthood' emerges as negative space in its world-building. As it homes, a nascent corporeality comes into being (returns?) through a thinking about an origin. The navel makes the body again.

Thus the body, as the previous home of thought, has returned in the corporeal productions of the new (sensing) machines of thought. This nozzle that feeds the making of a building—now inseparable from the 'machine' from which it 'takes-in' or does not 'take-in'; now the site of space's first differentiation; now the object of Filarete's autoerotic desire—like the navel that feeds Di Giorgio's city while oscillating between Cesarino's waist and pelvis, umbilicus and genitals, undoes the project of the body's erasure. Sensing flesh has returned, and with it error (and the fertile confusion of sin). In the homing of the machine we find simply the homing of the body, a relentless hauntology that will not go away.

Notes

1 Antonio di Piero (also known as Filaerete), 'Filarete's Treatise on Architecture: Being the Treatise by Antonio di Piero Averlino, Known as Filarete', trans. John R. Spencer (New Haven, CT: Yale University Press, 1965) Book I, Folio 4r, footnote 11, 8.

2 Spencer: "Filarete recognizes clearly one of the major problems of textual interpretation confronting any system of proportions based on Vitruvius. Any human figure drawn in strict conformity with Vitruvius' definition will have a different center when enclosed in a square from the same figure enclosed in a circle." Ibid, 8. See also Joseph Rykwert, *The Dancing Column: On Order in Architecture* (Cambridge: MIT Press, 1996).

3 These acrobatics were (to my knowledge) first and most brilliantly unpacked by Diana I. Agrest in *Architecture from Without: Theoretical Framings for a Critical Practice* at the beginning of the nineties—the last decade in which architectural discourse was really alive to the question of the body (Cambridge: MIT Press, 1991). The next year, Mark Wigley argued in 'Untitled: The Housing of Gender' that "It is not that the building is being thought of as a body with the classical analogy. Rather, the body is thought of as a building. The discourses of space and sexuality cannot be separated"; Beatriz Colomina, ed., *Sexuality and Space* (Cambridge: MIT Press, 1992), 326–389, 357. Throughout the ensuing decade Jennifer Bloomer continued to artfully confuse body and building and to parody the long tradition in which (male) architects cast buildings as their offspring. Some three decades later, I am indebted to my colleague Miguel Rodríguez-Casellas, for returning my thoughts to the question of the body, my own intellectual home.

4 Filarete, *Treatise on Architecture*, Book 1, folios 3vl-4rl, 8, my emphasis.

5 Amusingly, a tradition Le Corbusier continued when he shifted the navel coordinates between his 1951 and 1955 iterations of the *Modulor.*

6 A point only an Antipodean could make—I am indebted to my colleague Tim Schork for this point and for many witty conversations regarding robotic fabrication.

7 See Thomas Nagel, *The View from Nowhere* (Oxford: Oxford University Press, 1986).

8 See Ivan Sutherland, "Sketchpad: A man-machine graphical communication system," (PhD diss., University of Cambridge, 1963).

9 *MIT Science Reporter: Computer Sketchpad*, John Fitch, Dir. Russell Morash, WGBH-TV, Boston (21 mins.), 1964.

10 See Francesca Hughes, 'Deleted: How the Drawing that Can't Forget Forgot', *A/R/P/A Journal*, Columbia University Graduate School of Architecture, 'Instruments of Service' 04, 2 May 2016: http://www.arpajournal.net/deleted/. See also Leon Battista Alberti, *On Painting*, ed. Martin Kemp (London: Penguin Classics, 1991).
11 *MIT Science Reporter: Computer Sketchpad*, 1964.
12 1965 Gemini 4, *NASA's First Ever Space Walk*, Narrated by Ed White (June 3, 1965). https://www.youtube.com/watch?v=7K5DiKsZhTk.
13 *NASA Program Gemini Working Paper no. 5035*, NASA Manned Spacecraft Center, Houston, Texas, August 31, 1965, 97.
14 Francesco di Giorgio, *Il Vitruvio Magliabecchiano*, ed Gustina Scaglia (Florence: Gonelli, 1985) 28,r & v cited in Joseph Rykwert, *Dancing Column*, 64.
15 From Vitruvius's man and the propagative ambitions of Quatremère de Quincy's types, to Le Corbusier's *Modulor* and data generating 'actors' in optimizing flocking algorithms, the (male) body is deeply inscribed in architecture's harnessing of the (female) reproductive powers that have serially underpinned each new generative schema—be it predicated on the productive fertility of the line or the loop. This 'harnessing' is of course ancient: *Metis* represents not simply the name of the goddess of weaving (in every sense of the word) but also the attendant powers of transformative intelligence and transformative materialisation—both design and its delivery. Accordingly, Zeus swallowed *Metis* in the form of his pregnant wife, ingesting her shapeshifting fertility so that their progeny might not usurp him. See Anne Bergren, 'Dear Jennifer', 'Architecture and the Feminine: Mop-Up Work', *ANY*, 4 (January/February 1994), 12–15.
16 Agrest, *Architecture from Without*.
17 Filarete, *Treatise on Architecture*, Book II,7v, 15. My emphasis.
18 Filarete, ibid, Book II, 7v, 15, 16.
19 Filarete, ibid Book II, 7v, 16.
20 Mark Cousins, 'The New,' in *Friday Lectures*, ed. Parveen Adams (London:The Architectural Association, 2021) December 12 1997, 16.
21 Cousins, ibid, 14.
22 On the 'good enough mother' see Donald Winnicott, *Playing and Reality* (London: Routledge, 1991). Filarete, ibid Book II, 7v, 16.
23 Mark Cousins, 'The Anus and the Enclosure of Space', lecture delivered 10th November, 2000. AA/02/02/09/01/24/03, (90min), 4 digital files (604 MB). https://www.aaschool.ac.uk/markcousins/24
24 Cousins, Ibid.
25 Cousins, Ibid.
26 Cousins, Ibid.
27 He concludes the lecture with a typical coup de grace: this signals "not a question of truth and error but the emergence of a huge repression... The idea that architecture is constituted by enclosed space is an effect of a repression – it *is* the repression." Cousins, Ibid.
28 "What matters is not so much why the parents think shit is disgusting, of course they don't know. What they know is it is the disgusting object par excellence, and that therefore if the child is to be civilised, it is with the repression of this substance. It is entering the child into a world in which this is repressed that is fundamental." Cousins, Ibid.
29 Cousins, 'The New', *Friday Lectures*, 17, 18.
30 In my extreme anthropomorphic play here I am indebted to the extraordinary work of Madeline Gannon, and her robotic playmates, for making clear, lest there be any doubt, that when we are working on or with robots we are always working on or with ourselves.
31 Filarete, Ibid, book II, 8r, 16.
32 "When we say sexuality, what we mean is relations to the object." 'The New', *Friday Lectures*, 15.
33 Cousins, Ibid, 16.
34 Filarete, Ibid, book II, 8r, p16.
35 Ramon Llull's *Ars Genralis Ultima*, and its abbreviated version, the *Ars Brevis*, in a volume named the *Breviculum* is a kind of missionary artificial intelligence or logical system that sought to convert Muslims and Jews to Christianity. See Francesca Hughes, *Architectures of Prediction*, ARQ docs, ediciones ARQ, Santiago 2019.
36 As too was architecture's own autopoietic capacities. Aristotle sets out architecture's autopoietic capacity in *The Metaphysics*: architecture produces architecture. This was later manifest in the Enlightenment's type, a model that is able to infinitely generate architectural productions while protecting them from "vicious errors" (Quatremère de Quincy) in a process in which the type is not to be copied, but organises copying itself. While architecture's very much later embrace of cybernetics combines its own autopoietic legacy with Norbert Wiener's conflation of organism and machine. As the idea of a copying organism (as machine) is neatly transposed to that of the copying machine (as organism) all hangs on their ability to copy themselves, to reproduce without end, in a tunnel of infinite regress.
37 Greg Peters and Gavin Paul, "Maintaining an Old Icon with a New Technology", *Journal of Protective Coatings & Linings* 32 (2015): 22.

Bibliography

1965 Gemini 4. *NASA's First Ever Space Walk.* Narrated by Ed White. 1965. https://www.youtube.com/watch?v=7K5DiKsZhTk. Video.

Agrest, Diana. *Architecture from Without: Theoretical Framings.* Cambridge: MIT Press, 1993.

Bergren, Anne. 'Dear Jennifer.' 'Architecture and the Feminine: Mop-Up Work', *ANY,* 4 (January/February 1994).

Colomina, Beatriz, ed. *Sexuality and Space.* Cambridge: MIT Press, 1992.

Cousins, Mark. 'The New.' In *Friday Lectures.* ed. Parveen Adams. London: The Architectural Association, 2021.

Cousins, Mark. 'The Anus and the Enclosure of Space.' Lecture delivered 10 November, 2000. AA/02/02/09/01/24/03.

di Giorgio, Francesco. *Il Vitruvio Magliabecchiano.* Gustina Scaglia, ed. Florence: Gonelli, 1985.

di Piero, Antonio (also known as Filaerete). *Filarete's Treatise on Architecture: Being the Treatise.* Translated by John R. Spencer. New Haven, CT: Yale University Press, 1965.

Hughes, Francesca. 'Deleted: How the Drawing that Can't Forget Forgot.' *A/R/P/A Journal.* Columbia University Graduate School of Architecture, 'Instruments of Service' 04, 2 May 2016: http://www.arpa-journal.net/deleted/.

Hughes, Francesca. *Architectures of Prediction.* Santiago: ARQ docs, ediciones, 2019.

MIT Science Reporter: Computer Sketchpad, John Fitch, Dir. Russell Morash. 1965; Boston: WGBH-TV. Television.

NASA Program Gemini Working Paper no. 5035. NASA Manned Spacecraft Center. Houston, Texas. 1965.

Nagel, Thomas. *The View from Nowhere.* Oxford: Oxford University Press, 1986.

Peters Greg, and Gavin Paul. 'Maintaining an Old Icon with a New Technology.' *Journal of Protective Coatings & Linings* 32 (2015): 22–31.

Rykwert, Joseph. *The Dancing Column: On Order in Architecture.* Cambridge: MIT Press, 1996.

Sutherland, Ivan. 'Sketchpad: A man-machine graphical communication system.' PhD dissertation, University of Cambridge, 1963.

Winnicott, Donald. *Playing and Reality.* London: Routledge, 1991.

The Material

Prior to the Industrial Revolution, the notion of craft likely brought to mind the image of a skilled worker shaping some raw material into a singular, special object. For example, a cobbler pounding a sheet of leather to form the sole of a shoe or a blacksmith forging nails from wrought iron. Today, in our post-industrial and, arguably post-digital, world, the scenes of what some have begun to call "digital craft" are far different: gangly robotic arms weave together fibrous pavilions; fire-breathing laser cutters churn out ready-to-assemble kits of plastic parts; buzzing 3D printers materialize entire, custom objects ad infinitum from nothing but a single plastic thread. In such scenes, the craftsperson's interaction with the material they are crafting can be as fleeting as the whiff of the charred plastic or cardboard that emanates from the machine. In architecture at least, digital fabrication machines now largely serve as intermediaries between designers and their material workpieces, often putting us at varying distances from the things we make. How do these new modes of working with materials—i.e., those modes brought on by digital fabrication—influence how we craft our designs? How do they enhance or hinder our understanding of both the materials that we work with and the material bodies that we work in?

Central to this book's overarching proposition is the idea that the definition of digital fabrication can and does necessarily vary based on the practice of each individual digital fabricator. However, for the purposes of this section, we'd like to propose that digital fabrication can generally be described as the process of using computation to instruct machines to work on materials. The products of digital fabrication can thus be

DOI: 10.4324/9781003296522-9

thought of as material artifacts resulting from digitally-orchestrated processes. Conceived this way, digital fabrication can be seen as an interface between the immaterial digital environment that we now design in and the material world that we collectively inhabit. We then might wonder whether it's possible to use digital fabrication to imbue contemporary architectural designs with material traces of our digital design processes? If so, what are the aesthetics of these material traces? And what do these aesthetics signify for architectural culture, as well as material culture at large?

The prospect of more widespread digital fabrication in architectural design and construction processes carries cultural and material implications beyond the design studio. For instance, automated in situ construction will bear potentially dire consequences for building laborers; mass customization will bring about greater aesthetic diversity in the built environment; and digital reconstruction will augment the ways in which we restore and preserve buildings. On an ontological level, the progressively "intelligent" performances of digital fabrication machines will—and perhaps already have—distort common perceptions ideas of embodied intelligence and physical identity. We might use the work in the following section to forecast this future and ask ourselves: How does digital fabrication change our perception of what it means to be material?

The Ends of the Machine

From Technical Means to Architectural Beginnings

Leslie Lok and Sasa Zivkovic

On Means and Ends

Digital design and fabrication technologies have been intrinsic to HANNAH's work since we founded the studio in 2014. In our practice, we use such technologies to facilitate new material methods, tectonic articulations, environmental practices, technological affordances, and forms of construction. In our projects, we aim to mine the tension between machine *means* and architectural *ends* and reclaim authorship over processes of construction that influence the ways we currently build, or perhaps *ought* to build in the future. However, we try not to obsess over technology for technology's sake: as designers, we consider ourselves to be biased techno-generalists rather than techno-specialists. Our work's performances and architectural expressions emerge from its investigations into materiality, digital construction protocols, robotic routines, new technologies, unusual construction techniques, and bottom-up rule-based design logics. At the same time, we are motivated by top-down considerations, such as architectural precedents, select histories, ecological attitudes, collective labor, personal obsessions, and the creative misuse of technology. Ours is a design process defined by a mix of means.

The Oxford Dictionary defines "means" as "an instrument, agency, method, or course of action employed to attain some object or bring about some result."[1] While this definition of the word includes the notion of a "result"—or, end—we would argue that digital fabrication research in architecture is often disproportionately obsessed with its *means, methods, and tools* of production as opposed to the *ends* of such means.

DOI: 10.4324/9781003296522-10

This widespread obsession with technology's means often results in design research that remains tragically underdeveloped in scope, breadth, and real-world application.

Several of our peers have also noticed that the past 15 years of global production in materially driven digital fabrication research predominantly focused on[2,3] innovative means of making as its main driver and, sometimes, its singular end. Today, new robots, machines, end-effectors, software, and design processes continue to enable thousands of new iterations and sub-iterations of the ways in which architecture could be made. In fact, any tool imaginable (hammer, chisel, sander, jigsaw, wire cutter, printing nozzle, bandsaw, LED light, slingshot sling, rock dropper, toothpick, vegetable peeler, etc.) could be attached to a robotic arm, applied to almost any material, and translated to almost any scale using a variety of robotic fabrication procedures—architecture's myriad new means are truly endless. Simply pick a tool-machine-procedure combination and—if it hasn't already been explored—make sure to submit the resulting scientific paper of your work to the digital fabrication conference of your choice: you'll certainly receive serious consideration, if not acceptance.

As a result of this collective obsession for the means of digital fabrication, countless research labs—including our own, the Rural-Urban Building Innovation (RUBI) Lab and the Robotic Construction Laboratory (RCL)[4,5]—create endless prototypes, building methods, and small-scale "demonstrators," or, installations. Once published at a peer-reviewed conference or in a high-ranking journal, researchers often move on to invent the next new revolutionary means of future construction or marginally advance versions of their prior research. We are convinced that this rapid cycle of small-scale research production points to a paralyzed academic system that values and funds short-term publication success over wholistic long-term project development. In this research context, research hardly ever becomes thoroughly speculative, nor does it leave the lab and translate into architectural-scale projects, let alone buildings! Without follow through and/or scaling up, digital fabrication research inevitably remains a never-materializing promise, stuck in a self-perpetuating loop of its own continuous reinvention.

With few, but growing, exceptions,[6,7,8,9] digital fabrication research remains predominantly obsessed with its shiny new means of production and its associated low-hanging fruit, the proof-of-concept prototype. In an attempt to think past the end-of-the-research-road prototype and all the machine-means-informed research operations, we wonder: what are the ultimate aims—or the expanded ends—of all this work? Beyond academic "vanity" and promising prototypes, what is the endgame of the machine and its marvelous potentials for making?

On Beginnings

To discuss our own vision of the digital fabrication machine's endgame, we have to start with our beginnings in the field. In order to conduct research in advanced manufacturing and to engage new methods of making, one needs the facilities and tools—i.e., the means—to do such work. Despite working within a substantially privileged academic context at Cornell University, our research team did not immediately have access to large-scale digital fabrication tools or substantive startup funding. Nevertheless, we were profoundly interested in working on large-scale digital fabrication projects.

Access to state-of-the-art means of architectural fabrication continues to be a challenge, or outright impossibility, for those outside of well-endowed academic environments, architectural offices with considerable resources, or wealthy industrialized countries. Studies show that access to fabrication technology is also extremely uneven across the globe, skewing heavily toward industrialized regions.[10] At the same time, maker movements such as the *FabLab*[11,12,13] and the *RepRap*[14] have helped to make new means of production more broadly accessible. Such movements draw on the ethos of "*hacking*" and/or DIY[15] (do-it-yourself) sensibilities, as pioneered by software's open-source movement.[16] These relatively new approaches to technology can be useful strategies to significantly lower the initial startup costs for fabrication tools.

Adopting a DIY-mentality, the RCL at Cornell University developed the *Open-Source Factory* project[17] in 2016 and 2017. The project was a collaboration with teams of students in two consecutive architecture research options studios. The research consisted of the design, development, and documentation of two large-scale, open-source construction machines: a three-axis gantry 3D printer called *Daedalus* and an industrial robotic arm with a variety of end-effectors called *Dionysus*.

Daedalus is a scaled-up version of a desktop 3D printer that runs on the same readily available open-source software, as well as Arduino[18] firmware. The fully open-source machine has a print size of 9 × 18 × 9 feet and costs about $6,000 to make: a significant discount from a comparable gantry machine from an automation manufacturer.

In contrast, the *Dionysus* platform centers on a 20-year-old KUKA industrial 6-axis robotic arm, which was purchased on Ebay for $8,000 excluding shipping costs. A robotic arm is a thoroughly generalist tool, a characteristic that is highlighted by the fact that this particular robot welded cars for GM at a factory in Louisiana in its prior life. While not fully open source, due to proprietary KUKA components and hardware, Dionysus can be modified by attaching a range of end-effectors tools to enable a large variety of architectural research.

Both the printer and the robotic arm have enabled our research into digital fabrication's building-scale applications since our labs' founding. However, tools are constantly evolving and expanding, alongside with the discipline. Our investigation into architecture's new means of making has thus recently expanded to include Mixed Reality (MR) fabrication at the RUBI Laboratory at Cornell University.[19,20] This nascent research aims to enable "smart construction" without the use of heavy machines by using simple hardware and software-driven user-interfaces that provide complex assembly and construction instructions to their users. Here, heavy hardware is replaced by ethereal software environments.

From a design perspective, all of these new tools and software systems offer new sets of rules and constraints that designers can explore toward unique architectural ends. For instance, architects can now fully develop entire construction processes from the ground up, including the design of the construction tool or machine; in doing so, they can reclaim authorship over the process of construction itself.

The Beginning of the Ends

New machines and means of production, such as the robotic arm or the MR interface, can unquestionably influence how we design and think through making. However, in our opinion, successful architectural "fabrications"—what we define as the architectural imaginaries associated with making—require more nuance and finesse than mere investigations into assembly, efficiency, and novel construction techniques. Successful architectural fabrications should contain a vivid mix of complex narratives, histories, ethics, environmental considerations, contextual relationships, or design attitudes, among numerous other linkages. While the machine can provide the means for new forms of construction, it cannot help determine broader objectives—by itself, the machine offers no agenda and no agency. Further, while the mechanisms of the machine can help establish exciting new organizational and structural rulesets for design, they require designers who are willing to break those rulesets in order to create productive tensions and, ultimately, new fabrications.

At HANNAH, we are particularly interested in examining broader ends for three of our architectural fabrications: (a) design playfulness in concrete 3D printing, (b) environmental instrumentality in wood fabrication research, and (c) machine-derived tectonic expressionism at the architectural scale. In contrast to the more pragmatic and technical work we develop at our research labs, our work at HANNAH aims to expand architectural imaginaries more generally, exploring productive tensions between machine means and architectural ends, or vice versa. Those tensions often emerge from a careful balancing of the bottom-up rule sets of the machine with our own desire to challenge those very same rule sets in order to tell more complex architectural stories.

Design Playfulness and the Ends of Concrete 3D Printing

The development of reinforced concrete around 1850[21] enabled unprecedented architectural expression, afforded by unique material properties and structural technological advancements. In the nearly 170 years since reinforced concrete's invention, architects have made various claims about what characterizes an "architecture of concrete": heroic cantilevers, fluid forms, or distinct articulations of board form construction processes in exposed concrete structures.

The drainage scupper detail at the freestanding spiral staircase outside of the refectory at Le Corbusier's *La Tourette* monastery,[22] for example, is one of the most stunning architectural concrete details in one of the most remarkable concrete buildings ever built. The detail demonstrates a playful symbiosis between the means of material production and the ends of architectural design. Made from board-formed concrete, the scupper takes full advantage of the technical abilities of concrete as a material system, emphasizing and highlighting its plastic fluidity, its power to achieve dramatic cantilevers, its subtle wood-imprinted texture, and its comparatively easy formability. But the detail doesn't stop there: in a gentle formal gesture, the scupper is bent to align itself with the circular shape of the spiral staircase below in plan. In section, the scupper follows the

Figure 8.1 Our project *Additive Architectural Elements* aims to reveal the 3D printer's own idiosyncratic architectural tectonics.

gentle spiral dip of the roof geometry, fulfilling its functional requirement for efficient roof drainage in a delightful manner. Combining rational technical reason with expressive spatial intuition, Corbusier's spiral staircase drainage scupper detail is an amalgam of stunning architectural brilliance, material beauty, and profound intellectual satisfaction.[23] In many great Brutalist buildings, like the monastery, this oscillation between form, function, and process is expressively present.

Our work aims to introduce a kind of Brutalist balance to new processes of making, such as 3D printing concrete. The recent proliferation of building-scale concrete 3D printing technology is drastically altering past processes and means of concrete construction[24] to the point where we should now renew our discipline's age-old question and ask, "what is an architecture of 3D printed concrete"?

Our project, *Additive Architectural Elements – A New Robotic Brutalism*, aims to reveal the 3D printer's own idiosyncratic architectural tectonics and narratives (Figure 8.1). We chose to fabricate prototypical architectural motifs, such as floors, columns, doors, windows, walls, and ceilings, in order to explore how the layering and incremental corbelling of concrete can suggest new strategies for building (Figure 8.2).

In this process of re-imagining prototypical architectural elements through the printer, the manipulation of 3D printing rulesets and toolpaths became an opportunity for design. Each element in the series is described in one full-scale prototype, seven PLA models, 49 drawings, and one "architectural fantasy," in cube-form, which were derived from the elements' construction logics (Figure 8.3). While the research project includes full-scale speculative prototypes, it also serves as a conceptual and theoretical exploration into 3D printing, pairing bottom-up, technical aspects of the technology with top-down, formal design explorations.

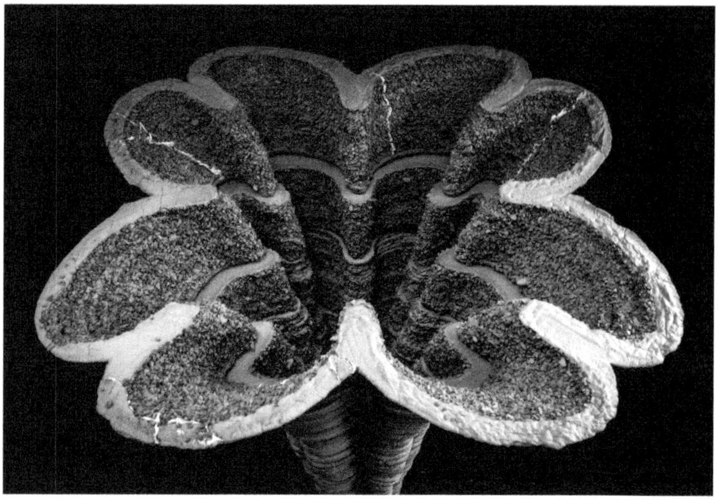

Figure 8.2 A flowering capital at the top of a 3D-printed corbelled concrete column.

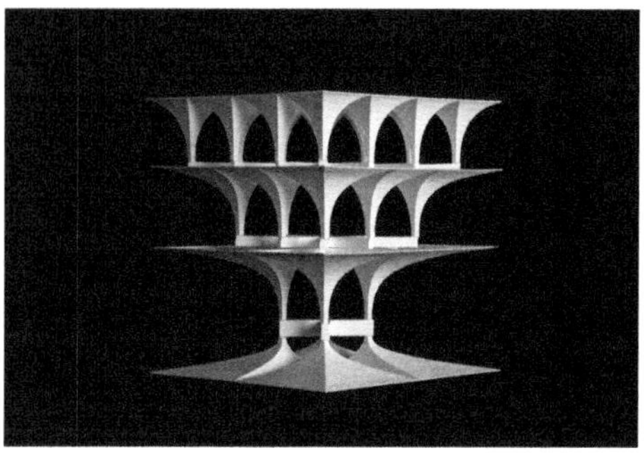

(a)

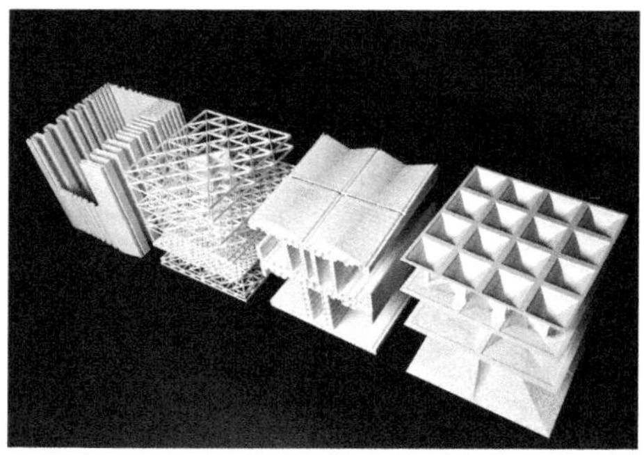

(b)

Figure 8.3 (a, b) The "architectural fantasies" derived from the construction logics of the *Additive Architectural Elements.*

The 3D printer—a Cartesian machine which has long been seen as characterless because of its wrongfully perceived ability to make any geometry—is in fact highly distinctive, and more complex than one might initially assume. As a particular type of 3D printer, the concrete 3D printer also exhibits distinctive nuance and personality. For example, concrete 3D printers cannot create rectangular window openings without human collaborators purposefully intervening in the manufacturing process; concrete, after all, cannot levitate in midair.[25] While the printer cannot print concrete in midair, the machine is very good at creating incremental cantilevers—or corbels—that can be used to generate complex spatial forms. For example, windows can be created by printing triangular corbelled arches (Figure 8.4). The *Corb. Window* element delineates how concrete 3D printers can create windows in line with their distinct functionality.

The *Doornament* element demonstrates how the manipulation of printing toolpaths into intricate lattice patterns can form wing-like gateways. Doors demarcate important, sometimes symbolic, thresholds into houses or buildings, and have historically

Figure 8.4 An incrementally cantilevered jamb in a 3D-printed concrete window opening.

been heavily ornamented for numerous purposes. The precise de-lamination of layers in the Doornament creates screen-like moments of transparency that at once highlight the toolpaths of the machine and generate an ornamentation reminiscent of Art Deco entryways (Figure 8.5). Beyond its visual impact, such ornamentation has the potential to become structurally performative,[26] and can be optimized to produce materially efficient components.

Concrete 3D printing requires the development of altogether new architectural languages that consider the limitations and inherent performative advantages of the fabrication process, as well as broader architectural imaginaries. The process' architectural languages can be productively enriched by individual architectural sensibilities and biases, be they technical, material, graphic, or formal. While the *Additive Architectural Elements* are highly influenced by the means and constraints of the machine, the machine never constitutes a design automatism but is rather a conversant and collaborator with whom we conceive new architectural tectonics.

Environmental Concerns and the Ends of the Bandsaw Robot

New means of architectural production enable architects to critically interrogate current construction practices and their environmental impacts. Today, over 90% of all new single-family homes in the United States are constructed with softwood dimensional

Figure 8.5 The delaminating layers in the *Doornament* create porous concrete screens.

lumber such as pine, spruce, and cedar from traditional sawmills.[27] Dimensional lumber has a lower carbon footprint than many other conventional building materials[28] but relies on non-local and fast-growing softwood from mono-cultural forests that don't encourage ecological diversity. Furthermore, the lumber for wood-framed buildings is often transported over great distances, originating from the Southern United States, the Pacific Northwest, or other countries. This long-distance transportation significantly increases the carbon footprint of the lumber industry.

As "optimized" dimensional lumber is shipped to construction sites across the world, local timber that has minor defects is dying in North American forests at monumental scales. The invasive Emerald Ash Borer (EAB) is a small beetle whose larvae hatch underneath a tree's bark, cutting off its nutrients by feeding on the cambium layer of native ash (*fraxinus americana*). The EAB threatens to eradicate most of the *8.7 billion* Ash trees in North America and has drastically transformed entire forest ecosystems in the process.[29,30] As of October 2018, it is found in 35 states and several Canadian provinces.[31] In New York State, where we live and where ash trees constitute about 8% of the overall tree population, the EAB was first discovered in 2009 and has since rapidly spread across the southern half of the state. While remediation programs exist,[32] they are often costly and the death of a vast number of Ash trees in North America is sadly inevitable. Infested Ash trees are locally available in many regions of North America but are rarely used in the construction industry despite retaining their structural properties for a period of up to two years after infestation.[33]

New logging protocols and fabrication methods can help utilize trees affected by the EAB epidemic. For example, HANNAH's *Ashen Cabin*[34] and RCL's *Log Knot*[35]

Figure 8.6 Our *Unlog* utilizes robotics and wood that's been infested by the EAB to reconsider standard timber framing practices.

deploy robotic fabrication and 3D scanning to repurpose irregularly shaped and EAB-infested timber for experimental construction. Irregular and curved tree geometries only make up for a small portion of the existing EAB-infested lumber, however. To generate impact at a larger scale, HANNAH's *Unlog* project works with predominantly mature and straight logs, with diameters below 12 inches and lengths between 8 and 10 feet (Figure 8.6). Such logs fall outside the standard parameters preferred by regional loggers, who harvest Ash for resale in lengths of 8, 10, 12, and 16 feet and a diameter above 12 inches. *Unlog* utilizes robotics and EAB-infested wood to broadly rethink logging practices, local supply chains, fabrication methodologies, and architectural tectonics in relation to standard practices of timber framing.

In *Unlog*, we developed a lightweight timber framing method utilizing kerfing techniques and a robotic arm that we equipped with a large-scale bandsaw (Figure 8.7). The project investigates how far a single EAB-infested log can be stretched, both literally as an assembly and figuratively as a resource (Figure 8.8).

In our robotic kerfing method, the robot makes successive incisions into a log, creating an accordion-like structural system that can expand and contract to create wall panels of varying density. We then assembled these panels into A-frames to create a spatial installation that loosely references framed structures, albeit in a much more materially efficient way. Oscillating between the precision of robotically cut surfaces and the natural geometry of the tree bark, *Unlog* creates an ever-transforming and undulating spatial assembly, illustrating the potentials of new tectonic strategies and sustainable construction methods (Figure 8.9).

Digital fabrication research in architecture needs to embrace environmental concerns as one of the main ends of our research means. New material practices, such as those demonstrated by *Unlog*, can begin to carve this new research trajectory.[36]

Figure 8.7 An industrial robotic arm equipped with a large-scale bandsaw machine the EAB-infested wood into lightweight timber components.

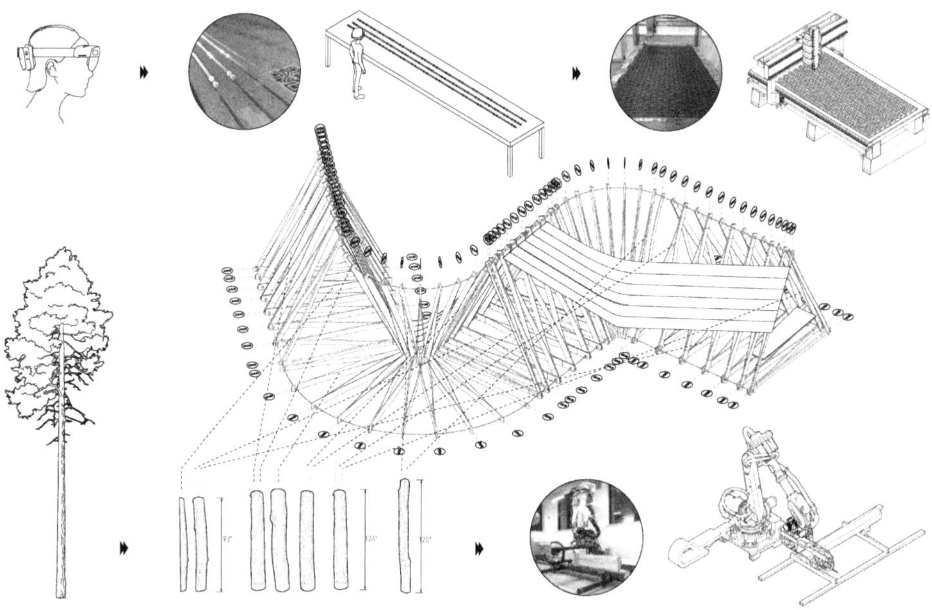

Figure 8.8 A diagram of the "stretched" *Unlog* assembly.

Moving forward, architects and researchers need to continue to develop new log-books, log new design protocols, and fundamentally un-log past loggings in order to move toward radically more sustainable and less wasteful building practices. At once rigorous and playful, *Unlog* leverages technological advancements to encourage a "nose-to-tail"

(a)

(b)

Figure 8.9 (a, b) *Unlog* demonstrates an accordion-like structural system composed of undulating A-frames.

usage of logs and timber that, by the harsh logics of our market economy, are broadly considered worthless and unusable.

Tectonic Expressions: Balancing Machine Means and Architectural Ends

Buildings are largely seen as the ultimate ends of architecture. As we mentioned earlier, there exist plenty of prototypes, pavilions, and installations that test and explore one specific aspect of a fabrication method, be it structural, aesthetic, or performative (our own work included). There are, however, far less projects that take fabrication methods and apply them, comprehensively from the ground up, at the scale of a building.[37] Our field's collective avoidance of the architectural scale, and all the complexity that comes with it, enables us to also avoid difficult questions that are essential to the architectural discipline, as well as the successful execution of any building; questions such as

> How does one design a door, a window, a floor? How does one integrate insulation, waterproofing, or an air and vapor barrier? How does the building's massing respond to site and context? How are spatial and circulation requirements met? How does one combine materials, craft details, or resolve the banalities of construction? How are building services integrated and resolved? How is building code addressed?

The list above could be expanded indefinitely and does not do justice to the complexity of even a simple architectural project; nevertheless, it hints at the multiplicity of relationships that architects have to address when creating buildings. Given the often enormous scope of architectural-scale construction, we believe it is critical to align new means and methods of construction with new ways of thinking about the design of architectural spaces.

Ashen Cabin is an attempt to synthesize bottom-up, research-driven digital fabrication prototypes with components of a building. The cabin is a small structure 3D-printed from concrete and clad in a robotically fabricated envelope made of irregular

Figure 8.10 An overall view of the *Ashen Cabin* perched in its landscape.

ash wood logs (Figure 8.10). The cabin has a footprint of 10 by 10 feet and is lifted off the ground by 3D-printed legs that are designed to work with the sloped terrain upon which they stand. All concrete components for the project were fabricated on *Daedalus*, our aforementioned self-built large-scale 3D printer, using a custom 3D printing process. The concrete structure comprises three programmatic elements: a table, a storage seat element, and a 25-foot-tall working fireplace. Expanding upon the *Additive Architectural Elements* project, the cabin reveals 3D printing's idiosyncratic tectonic language by exploring how the relentless three-dimensional deposition of extruded lines of concrete, when paired with the technique of corbelling, can suggest new strategies for building at full scale. Here, corbelling becomes an expressive and functional motif to highlight moments of programmatic significance throughout the concrete structure.

Ashen Cabin also challenges preconceived notions about material standards and environmental relationships in wood construction. Similar to *Unlog*, the Cabin utilizes curved wood infested by the EAB for its envelope. Irregular wood is largely unsuitable for construction due to its challenging geometries but can be sawn into naturally curved boards of varying thicknesses using a robotic arm that is outfitted with a special band saw (Figure 8.11). To create a façade envelope for the project, the boards are arrayed into interlocking Structural Insulated Panels (SIP) façade panels (Figure 8.12).

The SIPs are insulated using a two-component closed-cell spray foam (for which a fully biodegradable option is also available). The façade assembly is fully ventilated, detailed to manage shrinkage, and does not require an additional rain screen. No longer bound to the paradigm of industrial standardization (e.g., a 2×4), the resulting custom SIP panels revisit bygone wood craft and design based on organic, found, and living materials (Figure 8.13).

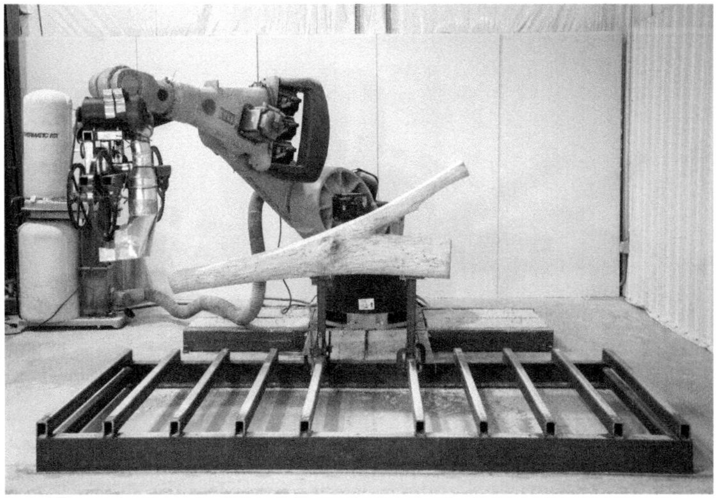

Figure 8.11 An irregularly shaped EAB-infested is set to be sliced into thin façade panels.

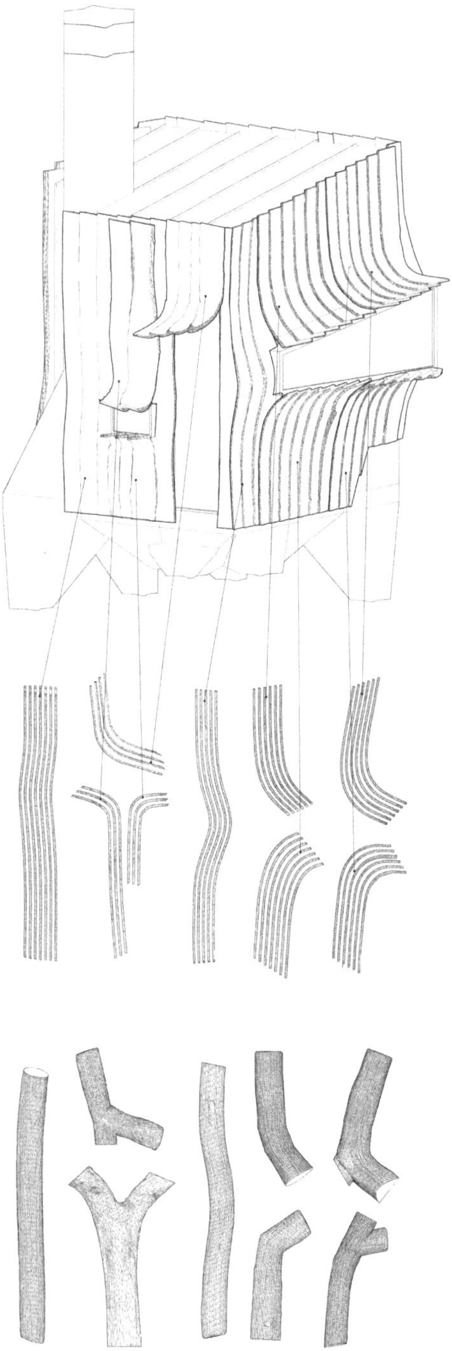

Figure 8.12 (a, b) The thinly sliced wood façade panels are interlocked to create a building envelope.

Figure 8.13 The "live edge" of the robotically band sawn wood panels.

(a) (b) (c)

Figure 8.14 (a, b, c) The curvature of the robotically band sawn wood drips is deployed to highlight moments of architectural importance.

Architecturally, Ashen Cabin walks the line between the familiar and the unfamiliar, as well as the technologically advanced and the formally elemental. The undulating wooden surfaces simultaneously embody their original, irregular geometries and accentuate the building's sculptural form. The curvature of the wood strips is strategically deployed to highlight moments of architectural importance, such as windows, entrances, scuppers, canopies, and door handles (Figure 8.14); it also provides programmatic opportunities, such as integrated shelving, desk space, or storage. On the exterior, bold black window geometries pierce the fabric-like wooden envelope (Figure 8.15). Over time, the exterior of the cabin will naturally gray, creating a subtler distinction between the two predominant material systems, concrete and wood.

Despite its highly articulated fabrication idiosyncrasies, the project rejects any exclusive dogmatism from machine-derived means; instead, it aims to combine a variety

Figure 8.15 The bold black windows pierce the Cabin's fabric-like wooden envelope.

of fabrication methods, material applications, geometries, and types of construction. Together, the various and sometimes conflicting means—robotic or otherwise—elucidate the architectural and tectonic expression unique to the project.

Means to an End: Toward New *Fabrications!*

Architecture's new technological means are powerful tools that can enable broad new endgames if they are developed and applied with purpose, for example, in buildings. The construction of buildings takes considerable amounts of economic resources and a colossal amount of time, which is often incompatible with the limited scope and funding of a lab research project. However, losing sight of the bigger, broader ends of architectural research into materially driven practices, i.e., buildings, is shortsighted and problematic. We need to find ways to re-infuse our current means of production with more expansive and comprehensive ends and not postpone critical disciplinary investigations to some unknown future date. We want to suggest that the building as an end of applied architectural research is a valid endeavor, especially in light of the climate crisis and a renewed focus on embodied carbon emissions for new building construction.[38]

The phrase "a means to an end" is defined as "any action, behavior, or object considered in terms of its results rather than in terms of its value, ethicality, etc., in and of itself; (also) the method by which any specific object is attained."[39] This expression is sometimes used to justify potentially questionable decision-making and, thus, can carry negative connotations. However, within our current environmental, social, and political environment, we are convinced of the need to collectively focus less on the *means* or methods of our research and more on its *ends*—the creation of new architectural fabrications requires it.

Notes

1 "mean, n.3". OED Online. June 2022. Oxford University Press. https://www-oed-com.proxy.library.cornell.edu/view/Entry/115436 (accessed June 11, 2022).
2 Slocum, Brian, Viola Ago, Shelby Doyle, Adam Marcus, Maria Yablonina, and Matias del Campo. "ACADIA 2020: Distributed Proximities." *Proceedings of the 40th Annual Conference of the Association for Computer Aided Design in Architecture*, Volume I: Technical Papers, Keynote Conversations. Acadia Publishing Company (2020).
3 Note: The authors acknowledge that ACADIA proceedings are broad and contain a multiplicity of different research trajectories. In this example, ACADIA, the North American flagship for computational design and fabrication research in architecture, is used as an example to illustrate biases in materially driven digital fabrication research, which predominantly focuses on process, methodology, and prototyping.
4 The RUBI Laboratory at Cornell University's Department of Architecture in the College of Architecture, Art and Planning (AAP) was founded by Assistant Professor Leslie Lok in 2018. RUBI Lab investigates novel forms of building and fabrication by coupling digital construction technologies with natural and non-standardized materials for the design of adaptable buildings and housing in rural-urban contexts. https://labs.aap.cornell.edu/rubi-lab.
5 The RCL at Cornell University's Department of Architecture in the College of Architecture, Art and Planning (AAP) was founded by Assistant Professor Sasa Zivkovic in 2016. The lab develops sustainable robotic construction technologies, material systems, and fabrication processes to advance the future built environment. https://labs.aap.cornell.edu/robotic-construction-lab
6 Schwinn, Tobias, and Achim Menges, "Fabrication agency: Landesgartenschau exhibition hall." *Architectural Design* 85, no. 5 (2015): 92–99.
7 Menges, Achim, Tobias Schwinn, and Oliver David Krieg. *Advancing Wood Architecture*. London: Taylor & Francis, 2016.
8 Dillenburger, Benjamin, Fabio Gramazio, Matthias Kohler, and Hannes Mayer. *How to Build A House*. ETH Zurich, 2020.
9 Philip F. Yuan. *Collaborative Laboratory: Works of Archi-Union and Fab-Union*. Hong Kong: Oscar Riera Ojeda Publishers. 2019.
10 Pick, James B., and Avijit Sarkar. *The global digital divides: Explaining change*. Berlin: Springer, 2015.
11 Mikhak, Bakhtiar, Christopher Lyon, Tim Gorton, Neil Gershenfeld, Caroline McEnnis, and Jason Taylor, "Fab Lab: An alternate model of ICT for development." In *2nd International Conference on Open Collaborative Design for Sustainable Innovation*, Vol. 17, pp. 1–7, 2002.
12 Walter-Herrmann, Julia, "Fablabs–A global social movement." *FabLab: Of Machines, Makers and Inventors* (2013): 33–46.
13 Blikstein, Paulo, "Digital fabrication and 'making' in education: The democratization of invention." *FabLabs: Of Machines, Makers and Inventors* 4, no. 1 (2013): 1–21.
14 Jones, Rhys, Patrick Haufe, Edward Sells, Pejman Iravani, Vik Olliver, Chris Palmer, and Adrian Bowyer, "RepRap–the replicating rapid prototyper." *Robotica* 29, no. 1 (2011): 177–191.
15 Spencer, Amy. *DIY: The rise of lo-fi culture*. Marion Boyars Publishers, 2008.
16 Lerner, Josh, and Jean Tirole, "The open source movement: Key research questions." *European Economic Review* 45, no. 4–6 (2001): 819–826.
17 Zivkovic, Sasa, and Christopher Battaglia, "Open source factory: Democratizing large-scale fabrication systems." In *ACADIA 2017: DISCIPLINES & DISRUPTION [Proceedings of the 37th Annual Conference of the Association for Computer Aided Design in Architecture (ACADIA) ISBN 978-0-692-96506-1] Cambridge, MA 2–4 November, 2017)*, pp. 660–669, 2017.
18 Banzi, Massimo, and Michael Shiloh, *Getting started with Arduino*. Sebastopol, CA: Maker Media, Inc., 2022.
19 Lok, Leslie, Asbiel Samaniago, and Lawson Spencer, "Timber De-standardized: A Mixed Reality Framework for the Assembly of Irregular Tree Log Structures". In *ACADIA 2021: Realignments toward Critical Computation*. [Paper Proceedings of the 41st Annual Conference of the Association for Computer Aided Design in Architecture (ACADIA)], 2021.
20 Lok, Leslie, and Jiyoon Bae, "Timber De-standardized 2.0: Mixed Reality Visualizations and User Interface for Processing Irregular Timber". In *25th International Conference on Computer-Aided Architectural Design Research in Asia, RE: Anthropocene, Design in the Age of Humans, CAADRIA*, 2022.
21 MacGregor, James Grierson, James K. Wight, Susanto Teng, and Paulus Irawan, *Reinforced Concrete: Mechanics and Design*. Vol. 3. Hoboken: NJ: Prentice Hall, 1997.
22 Potié, Philippe, "Le Corbusier. Le Couvent Sainte Marie de La Tourette/The Monastery of Sainte Marie de La Tourette." In *Le Corbusier. Le Couvent Sainte Marie de La Tourette/The Monastery of Sainte Marie de La Tourette*. Basel: Birkhäuser, 2015.
23 La Tourette is a particularly successful demonstration of the expressive potential of both board-formed and reinforced concrete. From a design perspective, the monastery is one of the early examples of brutalist architecture [29] which is interested in honest and expressive articulations of architecture's

material tectonics and functions [30]. Similar to the late work of Le Corbusier, the work of architects such as Alison and Peter Smithson, Paul Rudolph, or Ernő Goldfinger is deeply linked to the methods, processes, and technical abilities of cast-in-place concrete construction. In New Brutalism, the structural potential of concrete construction produces possibilities for new and bold formal expression which is informed by tectonic logics and material constraints. Paul Rudolph, for example, grooved his concrete to delay interior structural deterioration [31]. While there is little evidence that this particular process stops deterioration, there exists an ethos of implied reciprocity between process and form – often merged with a stunning design playfulness.

24 Wangler, Timothy, Nicolas Roussel, Freek P. Bos, Theo AM Salet, and Robert J. Flatt, "Digital concrete: A review." *Cement and Concrete Research* 123 (2019): 105780.

25 A 3D-printed rectangular window requires the insertion of a lintel that can temporarily support the window head until the material has sufficiently cured.

26 Battaglia, Christopher A., Martin Fields Miller, and Sasa Zivkovic, "Sub-additive 3D printing of optimized double curved concrete lattice structures." In *Robotic Fabrication in Architecture, Art and Design*. Cham: Springer, 242–255, 2018.

27 Dietz, Robert. 2019. "Framing Methods for Single-Family Homes: 2018: Eye On Housing." *Eye On Housing | National Association of Home Builders Discusses Economics and Housing Policy*, October 10, 2019. https://eyeonhousing.org/2019/10/framing-methods-for-single-family-homes-2018/.

28 Glover, Joanna, Donald O. White, and Timothy AG Langrish, "Wood versus concrete and steel in house construction: A life cycle assessment." *Journal of Forestry 100*, no. 8 (2002): 34–41.

29 Flower, Charles E., Kathleen S. Knight, and Miquel A. Gonzalez-Meler, "Impacts of the emerald ash borer (Agrilus planipennis Fairmaire) induced ash (Fraxinus spp.) mortality on forest carbon cycling and successional dynamics in the eastern United States." *Biological Invasions 15*, no. 4 (2013): 931–944.

30 Klooster, Wendy S., Kamal JK Gandhi, Lawrence C. Long, Kayla I. Perry, Kevin B. Rice, and Daniel A. Herms, "Ecological impacts of emerald ash borer in forests at the epicenter of the invasion in North America." *Forests 9*, no. 5 (2018): 250.

31 USDA Forest Service and Michigan State University, Emerald Ash Borer Information Network [online]. Available at: http://www.emeraldashborer.info (accessed 13 June 2022).

32 McCullough, Deborah G., Nathan W. Siegert, and John Bedford, "Slowing ash mortality: A potential strategy to slam emerald ash borer in outlier sites." In McManus, Katherine A; Gottschalk, Kurt W., eds. *Proceedings. 20th US Department of Agriculture Interagency Research Forum on Invasive Species 2009; 2009 January 13–16; Annapolis, MD. Gen. Tech. Rep. NRS-P-51*. Newtown Square, PA: US Department of Agriculture, Forest Service, Northern Research Station: 44–46, pp. 44–46, 2009.

33 Persad, Anand B., John Siefer, Roy Montan, Scott Kirby, Oscar J. Rocha, Michael E. Redding, Christopher M. Ranger, and Andrew W. Jones, "Effects of emerald ash borer infestation on the structure and material properties of ash trees." *Arboriculture and Urban Forestry 2013* 39, no. 1 (2013): 11–16.

34 Lok, Leslie, and Sasa Zivkovic, "Ashen Cabin". In *ACADIA 2020: Distributed Proximities. [Project Catalog of the 40th Annual Conference of the Association for Computer Aided Design in Architecture (ACADIA)]*, 2020.

35 Zivkovic, Sasa, Brian Havener, and Christopher Battaglia, "LOG KNOT: Robotically Fabricated Roundwood Timber Structure". In *ACADIA 2020: Distributed Proximities [Project Catalog of the 38th Annual Conference of the Association for Computer Aided Design in Architecture (ACADIA)]*, 2020.

36 Karl, Thomas R., and Kevin E. Trenberth, "Modern global climate change." *Science* 302, no. 5651 (2003): 1719–1723.

37 Note: See a select and rapidly growing list of exceptions in the introduction paragraph *On Means and Ends*.

38 Röck, Martin, Marcella Ruschi Mendes Saade, Maria Balouktsi, Freja Nygaard Rasmussen, Harpa Birgisdottir, Rolf Frischknecht, Guillaume Habert, Thomas Lützkendorf, and Alexander Passer, "Embodied GHG emissions of buildings–The hidden challenge for effective climate change mitigation." *Applied Energy* 258 (2020): 114107.

39 "means to an end". OED Online. January 2023. Oxford University Press. https://www.oed.com/view/Entry/115436?redirectedFrom=means+to+an+end#eid37448490 (accessed January 23, 2023).

Bibliography

Achim Menges, Tobias Schwinn, and Oliver David Krieg. *Advancing Wood Architecture*. London: Taylor & Francis, 2016.

Amy Spencer. *DIY: The Rise of Lo-fi Culture*. London: Marion Boyars Publishers, 2008.

Anand B. Persad, John Siefer, Roy Montan, Scott Kirby, Oscar J. Rocha, Michael E. Redding, Christopher M. Ranger, and Andrew W. Jones. "Effects of emerald ash borer infestation on the structure and material properties of ash trees." *Arboriculture and Urban Forestry 2013 39*, no. 1 (2013): 11–16.

Bakhtiar Mikhak, Christopher Lyon, Tim Gorton, Neil Gershenfeld, Caroline McEnnis, and Jason Taylor. "Fab Lab: An alternate model of ICT for development." *2nd International Conference on Open Collaborative Design for Sustainable Innovation*, Vol. 17, 2002, 1–7.

Benjamin Dillenburger, Fabio Gramazio, Matthias Kohler, and Hannes Mayer. *How to Build A House*. ETH Zurich, 2020.

Brian Slocum, Viola Ago, Shelby Doyle, Adam Marcus, Maria Yablonina, and Matias del Campo. "ACADIA 2020: Distributed Proximities." *Proceedings of the 40th Annual Conference of the Association for Computer Aided Design in Architecture*, Volume I: Technical Papers, Keynote Conversations, 2020. Mount Pleasant, SC: Acadia Publishing Company.

Charles E. Flower, Kathleen S. Knight, and Miquel A. Gonzalez-Meler. "Impacts of the emerald ash borer (Agrilus planipennis Fairmaire) induced ash (Fraxinus spp.) mortality on forest carbon cycling and successional dynamics in the eastern United States." *Biological Invasions 15*, no. 4 (2013): 931–944.

Christopher A. Battaglia, Martin Fields Miller, and Sasa Zivkovic. "Sub-Additive 3D Printing of Optimized Double Curved Concrete Lattice Structures." In *Robotic Fabrication in Architecture, Art and Design*, Zurich, CH, 2018, 242–255. Cham: Springer.

James B. Pick and Avijit Sarkar. *The Global Digital Divides: Explaining Change*. Berlin: Springer, 2015.

James Grierson Macgregor, James K. Wight, Susanto Teng, and Paulus Irawan. *Reinforced Concrete: Mechanics and Design*. Vol. 3. Hoboken, NJ: Prentice Hall, 1997.

Joanna Glover, Donald O. White, and Timothy AG Langrish. "Wood versus concrete and steel in house construction: A life cycle assessment." *Journal of Forestry 100*, no. 8 (2002): 34–41.

Josh Lerner and Jean Tirole. "The open source movement: Key research questions." *European Economic Review 45*, no. 4–6 (2001): 819–826.

Julia Walter-Herrmann. "Fablabs–A global social movement.' In Julia Walter-Herrmann and Corinne Büching, eds. *FabLab: Of Machines, Makers and Inventors*, 33–46. Bielefeld: transcript Verlag, 2013.

Leslie Lok, Asbiel Samaniago, and Lawson Spencer. "*Timber De-standardized: A Mixed Reality Framework for the Assembly of Irregular Tree Log Structures*". In *ACADIA 2021: Realignments Toward Critical Computation. Paper Proceedings of the 41st Annual Conference of the Association for Computer Aided Design in Architecture (ACADIA)], 2021.* Mount Pleasant, SC: Acadia Publishing Company.

Leslie Lok and Jiyoon Bae. *"Timber De-standardized 2.0: Mixed Reality Visualizations and User Interface for Processing Irregular Timber".* In *25th International Conference on Computer-Aided Architectural Design Research in Asia, RE: Anthropocene, Design in the Age of Humans, CAADRIA*, Sydney, AU, Sydney, 9–15 April 2022, 121–130. CAADRIA.

Leslie Lok and Sasa Zivkovic. "Ashen Cabin". In *ACADIA 2020: Distributed Proximities. [Project Catalog of the 40th Annual Conference of the Association for Computer Aided Design in Architecture (ACADIA)]*, 2020. Mount Pleasant, SC: Acadia Publishing Company.

Martin Röck, Marcella Ruschi Mendes Saade, Maria Balouktsi, Freja Nygaard Rasmussen, Harpa Birgisdottir, Rolf Frischknecht, Guillaume Habert, Thomas Lützkendorf, and Alexander Passer. "Embodied GHG emissions of buildings–The hidden challenge for effective climate change mitigation." *Applied Energy 258* (2020): 114107.

Massimo Banzi and Michael Shiloh. *Getting started with Arduino*. Sebastopol, CA: Maker Media, Inc., 2022.

McCullough, Deborah G., Nathan W. Siegert, and John Bedford. "Slowing ash mortality: A potential strategy to slam emerald ash borer in outlier sites." In McManus, Katherine A; Gottschalk, Kurt W., eds. *Proceedings. 20th US Department of Agriculture Interagency Research Forum on Invasive Species; 2009 January 13–16; Annapolis, MD. Gen. Tech. Rep. NRS-P-51*. 2009, 44–46. USDA.

"mean, n.3". OED Online. June 2022. Oxford University Press. https://www-oed-com.proxy.library.cornell.edu/view/Entry/115436 (accessed June 11, 2022).

"means to an end". OED Online. January 2023. Oxford University Press. https://www.oed.com/view/Entry/115436?redirectedFrom=means+to+an+end#eid37448490 (accessed January 23, 2023).

Paulo Blikstein. "Digital fabrication and 'making' in education: The democratization of invention." In Julia Walter-Herrmann and Corinne Büching, eds. *FabLab: Of Machines, Makers and Inventors*, 1–21. Bielefeld: transcript Verlag, 2013.

Philip F. Yuan. *Collaborative Laboratory: Works of Archi-Union and Fab-Union*. Hong Kong: Oscar Riera Ojeda Publishers, 2019.

Philippe, Potié. "Le Corbusier. Le Couvent Sainte Marie de La Tourette/The Monastery of Sainte Marie de La Tourette." In *Le Corbusier. Le Couvent Sainte Marie de La Tourette/The Monastery of Sainte Marie de La Tourette*. Basel: Birkhäuser, 2015.

Rhys Jones, Patrick Haufe, Edward Sells, Pejman Iravani, Vik Olliver, Chris Palmer, and Adrian Bowyer. "RepRap–the replicating rapid prototyper." *Robotica 29, no. 1*: 177–191 (2011).

Robert Dietz. "Framing methods for single-family homes: 2018: Eye on housing." *Eye On Housing | National Association of Home Builders Discusses Economics and Housing Policy*, October 10, 2019. https://eyeonhousing.org/2019/10/framing-methods-for-single-family-homes-2018/.

Sasa Zivkovic and Christopher Battaglia. "Open source factory: Democratizing large-scale fabrication systems." In *ACADIA 2017: DISCIPLINES & DISRUPTION, Proceedings of the 37th Annual Conference of the Association for Computer Aided Design in Architecture (ACADIA) ISBN 978-0-692-96506-1] Cambridge, MA*, 2–4 November, 2017, 660–669. Mount Pleasant, SC: Acadia Publishing Company.

Sasa Zivkovic, Brian Havener, and Christopher Battaglia, "LOG KNOT: Robotically Fabricated Roundwood Timber Structure". In *ACADIA 2020: Distributed Proximities [Project Catalog of the 38th Annual Conference of the Association for Computer Aided Design in Architecture (ACADIA)]*, 2020.

Thomas R. Karl, and Kevin E. Trenberth. "Modern global climate change." *Science 302, no. 5651* (2003): 1719–1723.

Timothy Wangler, Nicolas Roussel, Freek P. Bos, Theo AM Salet, and Robert J. Flatt. "Digital concrete: A review." *Cement and Concrete Research 123* (2019): 105780.

Tobias Schwinn and Achim Menges. "Fabrication agency: Landesgartenschau exhibition hall." *Architectural Design 85*, no. 5 (2015): 92–99.

USDA Forest Service and Michigan State University, Emerald Ash Borer Information Network [online]. Available at: http://www.emeraldashborer.info (accessed 13 June 2022).

Wendy S. Klooster, Kamal JK Gandhi, Lawrence C. Long, Kayla I. Perry, Kevin B. Rice, and Daniel A. Herms. "Ecological impacts of emerald ash borer in forests at the epicenter of the invasion in North America." *Forests 9*, no. 5 (2018): 250.

Working with Forests

Zachary Mollica

Introduction

People and trees have a long history together. We breathe thanks to the work of forests; most technological advances have relied in some way on trees and the products we derive from them; and in architecture and building, wood has been a central material to cultures across the world. Despite this rich history, the view dominating discourse in wood construction is basic: the more carbon we can 'lock up' in buildings, the better. This simplified goal fails to recognize that a standing tree both stores and continues to capture carbon, evades discussion of the immense impact industrial forest management has on soil health and biodiversity, and overlooks how energy-intensive most uses of wood in construction are. New digital tools, which can provide solutions to the problems their mechanical predecessors created, too often only exacerbate the same problems (Figure 9.1).

The cutting edge in wood building now is cross-laminated timber and other engineered wood products. But the creation of these demands that trees be cut into countless small pieces, dried, and glued or otherwise fastened back together to make mass products. This process privileges the pursuit of calculable, consistent material, and regards non-conforming trees with 'defects' as near useless. When buildings are made from a few species and shapes of trees, the conforming varieties are the ones that are encouraged to grow. This approach to using wood radically transforms 'forest forms'[1] and removes the rich biodiversity that produces healthy forests. As a result, forests around the world have started to bear a kind of self-similarity akin to that of the suburbs.

DOI: 10.4324/9781003296522-11

Figure 9.1 3D scanning and building in collaboration with the forest.

This case study explores alternative approaches to working with the products of forests in ways that promote biodiversity, healthy and economic forest management, and ecologically responsible design; further, it puts forth methods for designing with the form and best properties of diverse trees and forests. Here, digital surveying tools such as 3D scanning are presented as instruments that allow us to carefully observe and document trees with the goal of deploying their products more intentionally in buildings. By re-diversifying wood construction methods, the process of making buildings can be made an active contributor in the regeneration and expansion of thriving forest ecosystems.

The trajectory of work presented has evolved from the capture of isolated tree elements to comprehensive models of whole-standing trees, to ongoing work on large-scale detailed forest scan inventories (Figure 9.2). The goal of this work is not to digitally automate wood construction from photosynthesis to finished form; instead, it is to attempt sophisticated combinations of digital and traditional processes that, together with human input, can establish locally adaptable methods for distributed manufacturing.

How We Manage and Use Trees Now

Forestry as it is practiced across much of the Western world today is based on growing straight, cylindrical segments of tree ready to extract and ship globally. To increase profitability, the industry view is that these lengths should be as straight as possible, be between the minimum and maximum diameter of an available sawmill, and possess few knots. Trees fitting this description are a small portion of those within a 'natural' forest condition, and so over time, methods of management have developed to encourage uniformity in forests (Figure 9.3).

Figure 9.2 Elevational view extracted from a large 3D scan model depicting five acres of mixed species forest.

Figure 9.3 Eight among many moments in the process of growing a managed conifer tree compartment. Left to right from top left: (1) small saplings in a box; (2) saplings planted closely together; (3) young trees begin to crowd into each other; (4) result of a first thinning and low branch pruning; (5) a second round of thinning trees; (6) a relatively matured stand near to its final form; (7) conifer forest after clear-felling; (8) a portion of the roundwood logs harvested. *Images do not depict a single forest area.

As an example, look at conifer forests. Typically faster growing than deciduous trees, conifer species such as spruce and fir contribute much of the volume of wood used in construction. To establish a managed conifer forest, young saplings are planted following a tight grid of maybe four feet by four feet.[2] By planting closely, trees are encouraged to grow straight and rapidly upward in competition for the sun's rays. As they begin to contact each other's branches, a forester will begin a cyclical effort known as 'thinning', by which trees are selected and cut down, their removal allowing remaining trees to amass volume rapidly. When a compartment is ready for final harvest, the trees standing may represent a fifth or less of those planted.

There are many ecological impacts from growing trees in this hurried manner, but a significant one to note is that trees grown this way will tend to have a less hearty rooting to the ground. This leaves them prone to blowing over and means they can easily be viewed as a liability rather than an asset the longer they remain standing. This is one reason that these trees are likely to be clear-felled and the whole process started over.

A Fork in the Road

In the 1980s, Parnham College demonstrated a new way of working with and managing trees at their 350-acre forest site known as Hooke Park in Dorset, UK. This forest was entirely clear-felled during post-WWII rebuilding efforts and was replanted with a range of species-segregated tree compartments in the '50s. It was then left unmanaged until the '80s when John Makepeace, the College's director, fortuitously saw that designers could add value to 'low-value'[3] forest products and, in doing so, promote better forest stewardship.

Forestry is a hard business, with large volumes, long wait times, and small margins. By adding value to the byproducts of forest management – itself a costly effort – Makepeace saw that foresters could receive greater income throughout the life of a compartment. Three nonstandard buildings developed in Hooke Park in this era demonstrated its ethos.[4] For each, the project teams would set their eyes on adding value to the long, small diameter 'thinnings' previously discussed. Too slender to be milled, the engineers concluded the long poles could be made to work ambitiously if loaded axially. The workshop building shown in Figure 9.4 demonstrates this approach with a series of arches spanning nearly 15 m.

From 2014 to 2021, Hooke Park was my home. In 2015, I led a team in the design and production of the 'Tree Fork Truss' (Figure 9.5) as part of the Architectural Association's MArch Design + Make program.[5,6] The project was instigated and advised by the program's founding director, Martin Self. Supplementing Parnham's approach to using forest products with new digital tools, the project developed a then-unprecedented workflow for employing tree forks typically disregarded by industrial practices. The project shows how a non-linear tree form can be observed, extracted, processed, and installed as a structural element.

Figure 9.4 Hooke Park workshop. A large spanning building constructed in and of the forest.

Figure 9.5 The Tree Fork Truss' primary structure assembled on site in Hooke Park.

Finding and Extracting Forms in the Forest

To design with the forms of trees requires knowing them intimately. The Tree Fork Truss took cues from traditional uses of non-linear wooden members by boatbuilders (Figure 9.6).

Figure 9.6 A plate from the Encyclopédie Méthodique (Plates, Vol. 1) shows examples of 'compass timbers', non-linear tree segments that were sought out by traditional craftspeople for the irregular grain patterns they possess.

Using a curved piece of tree, whose grain matches the curve of the component being made, builders could produce boats that were both stronger and more flexible. To find these 'forest forms', boatbuilders traversed local forests with templates of the shapes needed, using these to visually identify appropriately shaped tree pieces in the canopy. Starting in 2D, our own process adapted this method with the use of iOS app *Theodolite* to gather photos of 204 forks spread across Hooke Park (Figure 9.7). Photographs captured from the ground are inevitably distorted when documenting forms in the canopy of a tree, and so a short script was developed to correct them.[7] The base of each tree was physically marked and had its GPS location recorded so that it could be found again.

With these corrected and scaled drawings of each fork, it became possible to analyze and compare the forms available in the forest. While 204 forks identified within 100 acres made up only a tiny fraction of the non-linear forms present, this database did enable us to make some conclusions. First, we identified a relationship between the diameter of a fork and its angle of opening: as forks become larger, they tend to narrow in their opening angle to resist their cantilever failing. Where small twigs might be commonly found to open more than 45°, at approximately 150–200 mm diameter, the average beech tree fork in Hooke Park opens just 24°. These values became critical to drawing the structure we imagined (Figure 9.8).

A cutting list was made of 40 forks with relatively consistent diameters and the most open angles we could find. From this list, 25 were selected and felled, with several forks removed from the list in consultation with the managing forester. By harvesting the forks ourselves, we were able to draw a key conclusion: no matter how well we documented the forks, they can easily break if a tree is brought down in the wrong direction. To avoid this, great effort was given to felling trees in such a way that the forks landed flat (Figure 9.9).

Figure 9.7 An image captured in the Theodolite iOS app shows one of the forks surveyed in the Tree Fork Truss project alongside photo metadata and other information that would be recorded.

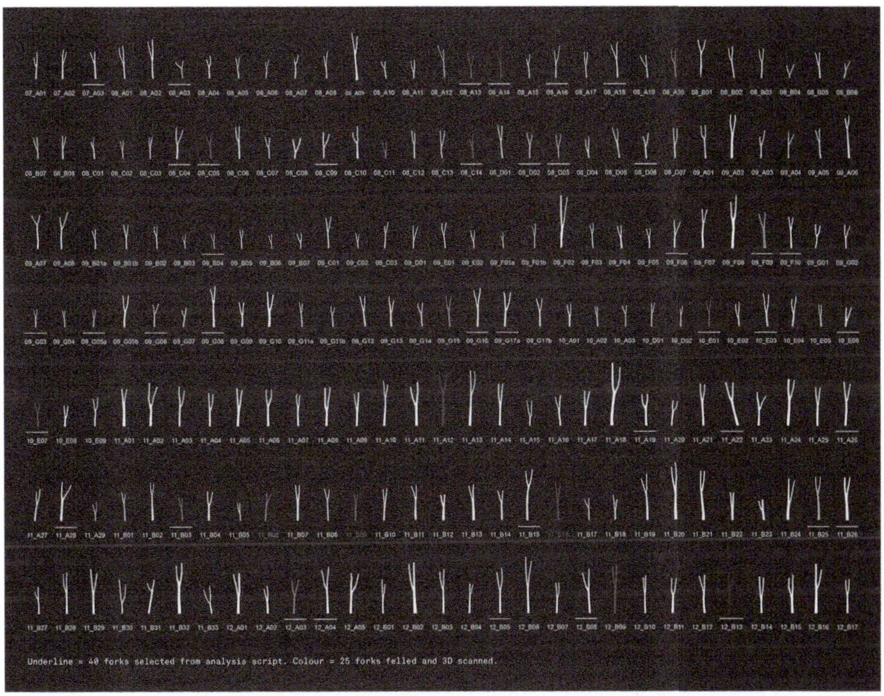

Figure 9.8 Diagram depicting 204 forked forms that were documented across Hooke Park's beech compartments. In this database, it was possible to evaluate the whole stock together against conditions such as: 'forks whose limbs open more than 35°'. Underlined are the 40 selected for review, and colored are those that were felled and used in the project.

Figure 9.9 Christopher Sadd, Hooke Park's Head Forester, inspecting a tree fork broken during felling. This occurred both due to the angle the tree hit the ground, and due to a small area of internal rot only apparent after the breakage.

3D Scanning Tree Forms in Order to Use Them

To compose and detail our structure, and to begin fabrication, we needed a more detailed digital drawing of each fork. After testing several handheld 3D scanning devices, we identified photogrammetry as the best tool for documenting large tree elements. This digital process uses numerous still photographs of an object to construct a detailed 3D model either in point cloud or mesh form. Requiring off-the-shelf hardware such as an iPhone or SLR camera in place of specialist kit, this process represents both a potent and accessible method of 3D scanning.

An average of 60 photos were taken to wholly capture a fork 6 m long. These photos were taken all around the object (Figure 9.10), with sufficient overlap between them to enable correlation by the software used.[8] Scanning any object through photogrammetry presents new considerations. For example, to accurately capture all faces of a fork, it was critical to mitigate the presence of highlights/shadows created when photographing toward the sky. Photogrammetric methods are not correctly to scale by default. This added a further requirement for us to create a physical scaling device using a known measurement on each fork to correlate digital representation and physical object. Once the scanning of each fork was complete, 3D mesh models of 25 forks were exported to be worked with in Rhinoceros and Grasshopper.

A common challenge in 3D scanning work is that it is easy to capture masses of detailed data (terabytes even), but far harder to leverage the most salient information efficiently. Each fork scan was made up of nearly 100,000 triangular faces. To get around the challenge of precisely orienting these models, we developed a rationalization tool

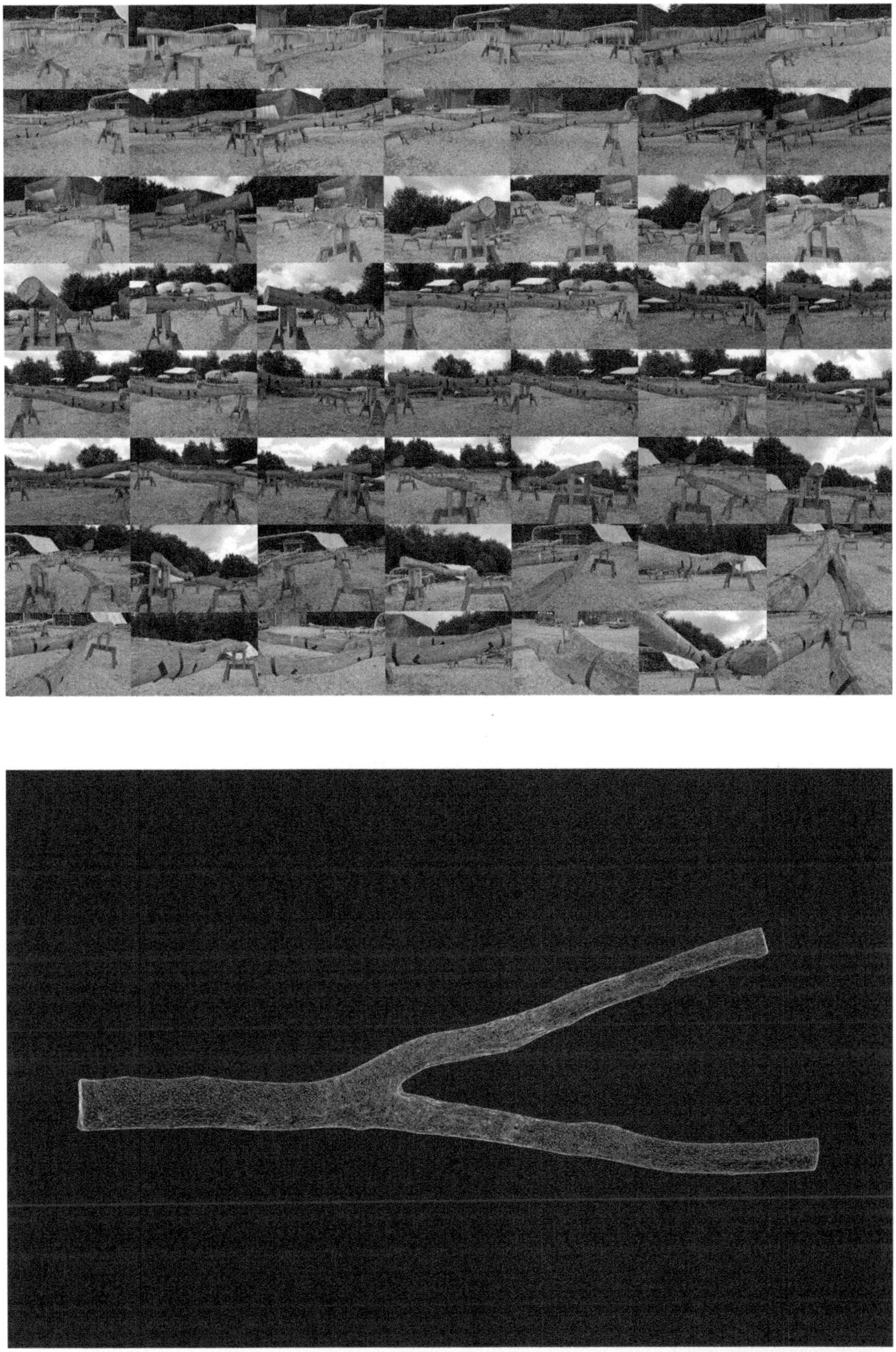

Figure 9.10 Numerous photographs captured to produce a high-resolution 3D scan of a fork through photogrammetry and a slide showing the cleaned-up mesh file produced.

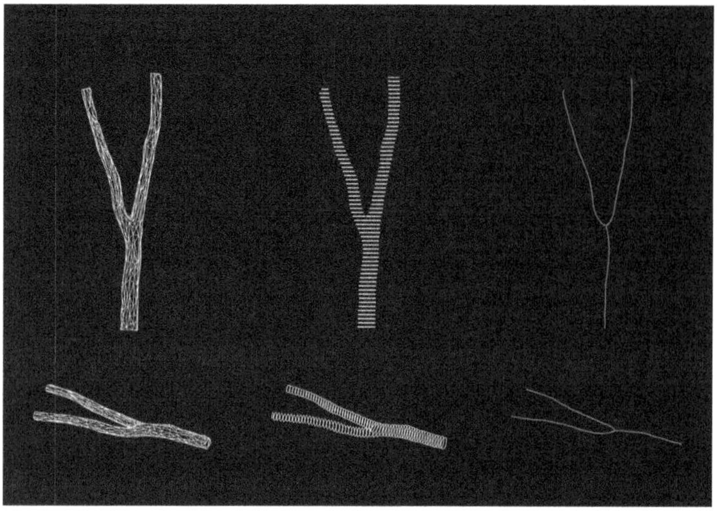

Figure 9.11 A scripted procedure was developed to produce centerline representations of each fork. The process was based on cutting many contour sections through the round material, estimating the center of these resulting curves, and then connecting the centers together in the correct order.

that simplified each fork to a set of three centerline curves (Figure 9.11). These curves, and their related scans, could then be precisely manipulated in CAD software.[9]

Scanning the Rest of a Tree

The Tree Fork Truss initially results in a paradox. It is premised on the increased value that can be ascribed to a fork through digital methods, yet each fork required the felling of an entire tree without plans for high-value uses for the rest of its mass. This observation, as well as an intention to demonstrate the potential value of this work beyond the scale of the one-off build, have propelled me through a series of increasingly detailed projects aiming to 3D scan trees in their entirety.

I set out an initial experiment to test the performance of different photogrammetry software in 3D scanning parts of trees that were more unusual in their natural forms than forks. Several scans were completed of the root plate of an old oak tree (Figure 9.12) which had been felled in the 1940s but was left upside down in the forest, deemed worthless at the time. Parts of a tree that grow underground have unique grain patterns owing to the sharp bends of their roots and tend to be particularly resistant to rot, having spent most of their life underground. Through this comparative technological testing, Agisoft's Metashape Pro software emerged as the primary tool for my continuing photogrammetry research, providing nuanced controls and ready troubleshooting.

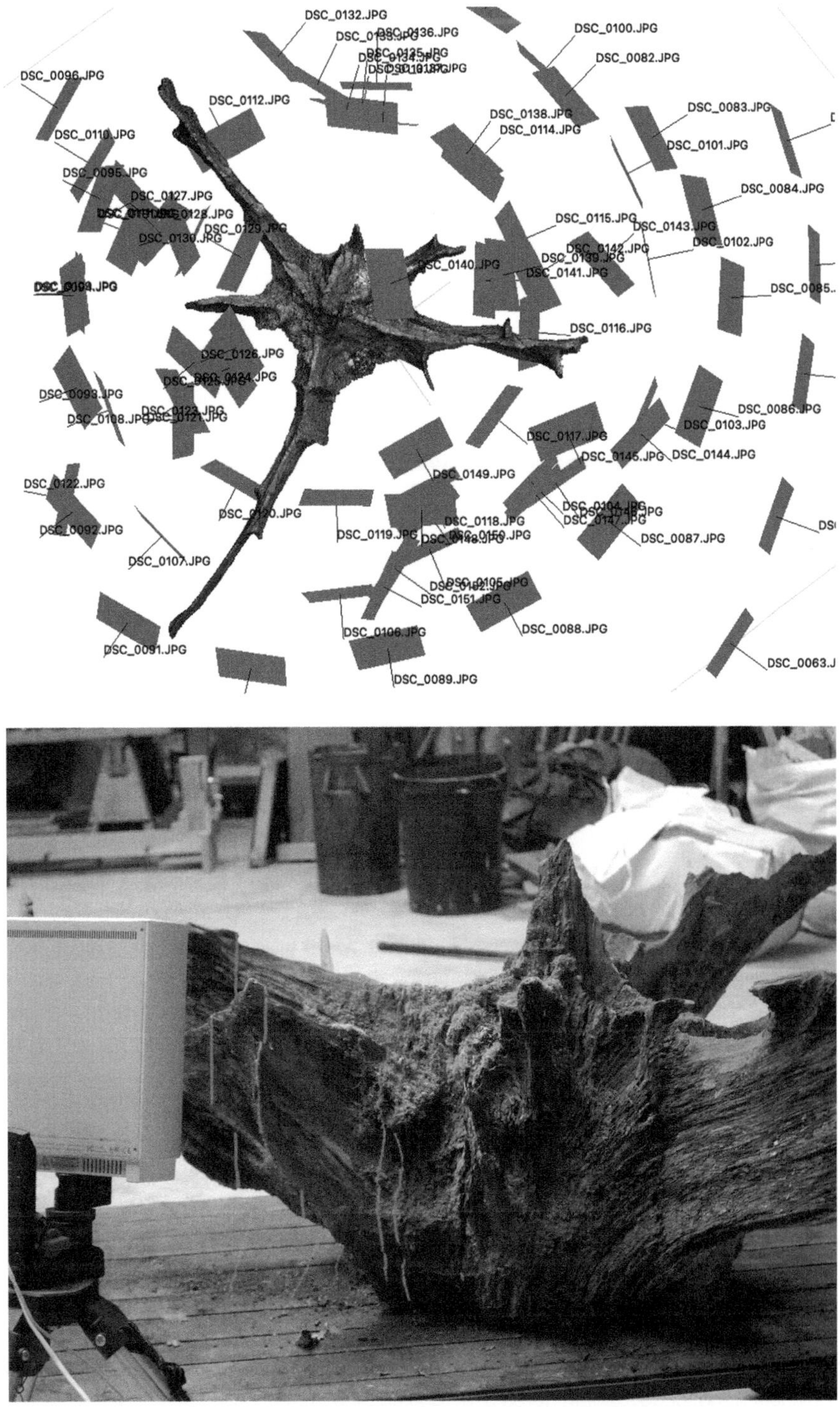

Figure 9.12 3D scan of the root plate of a large oak tree left lying in the forest in the 1940s.

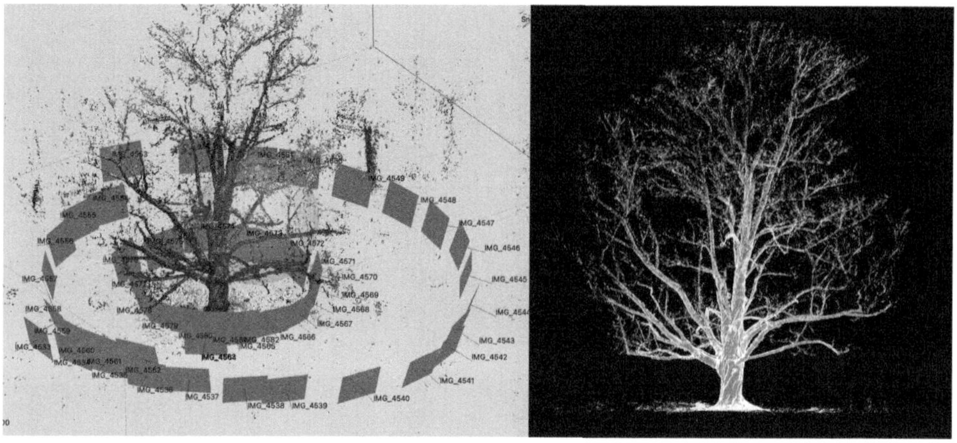

Figure 9.13 Left: In progress photogrammetry model of 'Mother Maple', an elder tree in the middle of Windhorse Farms, Nova Scotia. Oriented around a low-resolution point cloud, each blue rectangle Identifies to location and orientation of one of the images captured in 3D scanning. Right: Render from a completed scan of 'Mother Maple'.

With greater experience and a more capable toolset, I began to scale up toward scanning whole standing trees to study them better. Figure 9.13 depicts an elder maple – at least 300 years old – that was 3D scanned from the ground to its canopy with just 60 photos. Photos were captured in a spiral around the tree's base and were aligned to provide an accurate model of the whole. Close inspection revealed that the 3D data was less accurate the higher you looked in the tree—its branches fading into disconnected clusters of points. This was in part due to these branches being quite slender, and therefore difficult to photograph against the stark backdrop of a bright sky. But a few additional factors were evident. The photos we used had all been taken from the ground, and so due to the distance from the camera's sensor, there was a loss of resolution in images the higher we looked up. We tested capturing additional photos with the help of a drone, but a persistent issue remained: the canopy itself moves in the wind, even on the most still day.

Scanning Whole Forests

By 3D scanning trees and parts of them, the work discussed here demonstrates ways to observe, process, and put into use a more diverse range of 'forest forms' than is common in traditional industrial loggin practices. At the same time, we have been continually reminded by new scholarship that trees are collective entities and that we cannot expect to understand or to benefit forests by considering them individually, no matter the detail.[10] We have learned how trees communicate underground, the outsized role elder trees play in an ecosystem, and critically, the importance of factoring soil health into any accurate measurement of carbon sequestered in both forests and buildings.

To contribute to this emerging space, my latest work is oriented toward the capture and analysis of 3D scans representing increasingly large tracts of forest. In these ongoing projects, I seek to find ways in which the same tools of 'close observation' can help us to not only use tree components but also to understand which trees may be best to remove, and in doing so, how neighboring trees will be affected.

In this scaling up, my 3D scanning work has come to rely more often on capturing data through Light Detection and Ranging (LiDAR).[11] This approach utilizes specialist hardware that combines precise measurement of the time taken for a laser pulse to return along a trajectory together with spherical photography to construct enormously detailed, full-color representations of objects and environments. Covering an area of approximately five acres, the 3D data of my latest work shown in Figure 9.14 contains geometric information on approximately 2,000 trees of varied ages and species (including, but not limited to, ash, spruce, fir, oak, and cedar as well as small vegetation). The model representing them includes information on tree shapes, canopy extents, and terrain, and has been compiled through the combination of 100 discrete LiDAR scans.

The critical task in working with this data is to begin to give it structure. As I mentioned before, to have two billion points is not itself useful, and so a few first steps must be undertaken. Through segmentation, the point cloud data is separated into ground and above ground. The ground readily provides topography lines, while further segmentations can be made to separate small and large vegetation and to separate the canopy from the trunks of trees. Simultaneously, the point density of certain areas is greatly reduced through sub-sampling procedures in order to maintain a workable file size.

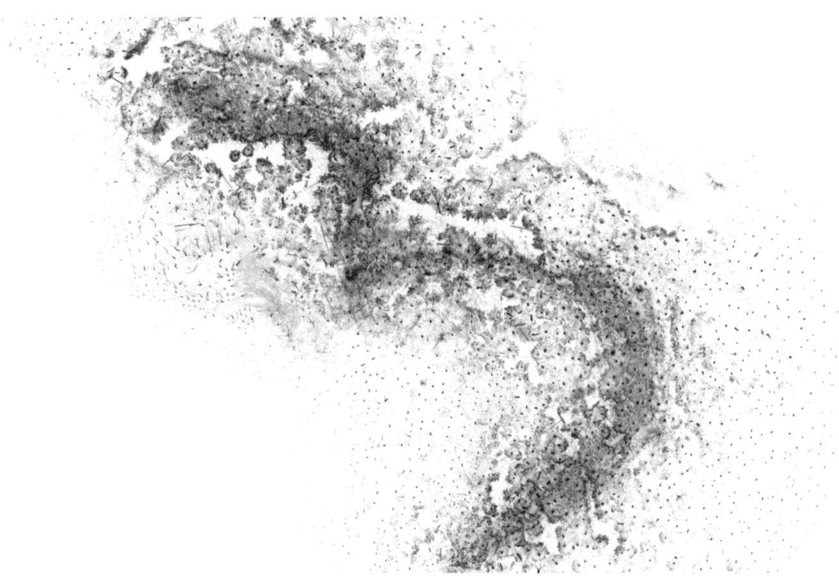

Figure 9.14 Plan view drawing created from a five-acre 3D scan captured through LiDAR. Each spider-like cluster in the view is a whole tree with approximately 2000 present across the whole image.

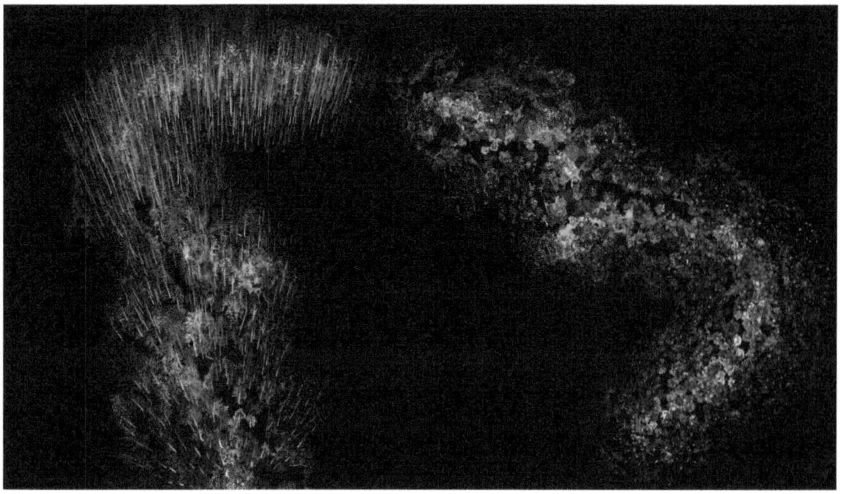

Figure 9.15 A view showing the segmentation of a forest point cloud into individual tree point clouds. Tones are arbitrarily assigned but denotes that this data is now identifiable separate to any neighboring trees and ready for detailed analysis.

My recent interrogations of this dataset seek to analyze and measure the overall 'forest form' for several purposes. First, from this detailed modeling, I can derive accurate estimates of the woody mass (carbon sequestered) in a particular area. Unlike traditional methods for estimating forest volume, which act on rules of thumb and samplings, these calculations can include the variation present in every individual tree's form. This same model – a point cloud given solidity through voxelization – will over time be able to provide a platform to conduct wind and solar analysis on the stand's trees, identifying, for example, the value or harm the removal of a particular tree might have (Figure 9.15).

The second major area of this work has been in automating the process of identifying forked and otherwise bent forms in the forest that might be useable. These large forest scan models present the opportunity to approach forest compartments as standing material libraries, and the quality of LiDAR data increasingly provides the capability to identify tree forms that might not be easily spotted by eye from ground level. The process begins by further segmenting the forest so that each tree is classified individually (Figure 9.16). With each tree distinguished, larger scale centerlining algorithms can be run to trace simplified line representations into the point cloud of each. From there, the search begins for unusual pieces of tree of value.

Scaling this Work Further

> We have to un-blind ourselves to human exceptionalism. That's the real challenge. Unless forest-health is our health, we're never going to get beyond appetite as a motivator in the world[12]

Figure 9.16 Stud-A, a small house in and of the forest, was designed and built through 2021 by students of mine in Hooke Park. Given a brief, 'How We Make Houses with Trees', the students demonstrated the use of a combination of alternative forest products: round 'thinnings' to form a shear wall, tiny ash coppice to make a new wooden stud, cedar bark transformed into shingles, and sawn green wood filling out the rest. Images from the students.[13]

Ways of working with wood that are as diverse as the species of trees we know and the regions they are local to are badly needed. To achieve such radical (rather than incremental) change demands that we establish new methods and priorities. Beyond making manufacturing processes faster, digital technologies can enhance our knowledge of trees and, ultimately, help us develop more ecological ways of working with them. This is a fork in the road that has not yet been taken.

In the near term, 3D scanning and related digital toolsets present means to engage the world's forests in the shape they are now in. Though this research prioritizes information gathering rather than physical fabrication, it suggests that digital fabrication is not simply about working with materials but also about developing intimate knowledge of these materials. In the case of trees, by enabling the use of a wider range of forest products, methods of close observation can over time enable the restoration of wilder forest forms across the world without discouraging the cutting of trees. There is a lot to do to take this work into the world at scale, and each step will require the effort of many collaborators.

A few observations made in all this time looking at wood, trees, and forests, guide my current thinking:

- The gap between the experimental wood fabrication research conducted at educational institutions and the reality of wood construction practices out in the world is vast. We need to prioritize making tools that enable non-specialists to deploy diverse forest products in their own work.

- Trees grow on a long timescale, requiring forest managers to foresee valuable markets 70 years away. Tools of 'close observation' allow us to work efficiently with trees exactly as they are now.
- The actual shape of a tree is elusive: they flex notably in the wind, transform through the seasons, the wood we take from trees changes shape constantly as it dries, and the whole form of a tree is transformed once felled – its limbs released from the weight of gravity.
- A goal that should be stated more explicitly is to actively encourage the use of 'wet wood'. If we can lessen just kiln drying in our production of wood from trees, we will make significant progress.
- New methods for in-forest extraction of trees that are both sensitive and scalable are needed. Neither a return to working horses or swarms of drones is likely to be the solution.
- Just enough technology and power should be used to get the work we are pursuing done. The short story 'On Exactitude [or Rigor] in Science' is an excellent reminder of where we otherwise head and is particularly relevant to any 3D scanning work.

Notes

1 'Forest Form' and similar terms are used in this work to variously describe the shape of individual tree elements such as forks, the overall structure of standing trees, and the arrangement/physical features of trees that viewed together make up forests. In general, it is used to identify the inherent complexity of natural conditions in contrast to more uniformed trees encouraged by human intervention.

2 In these paragraphs and those following, there is no absolute right answer on the methods or figures used. Each forest and forester tend to have their own. Details provided are therefore from my own experience gained in collaboration with practitioners. The principles they represent are more critical than the numbers.

3 Phrases 'low-value' and 'high-value' are used typically in this work to convey a sense of the conventional economic value that can be assigned to these trees and their parts. These simplified terms ignore many other characteristics and values that trees present and should not be read to mean simply good or bad trees.

4 Prototype House (1986) and Workshop (1989): ABK Architects, Buro Happold, Frei Otto, and others. Westminster Lodge (1997): Edward Cullinan Architects, Buro Happold, and others. Each project a deep collaboration between architects, engineers, craftspeople, and foresters.

5 In 2002, Parnham College transferred ownership of Hooke Park to the Architectural Association School of Architecture. From 2014, I completed the AA's 16-month MArch Design + Make program then became a member of AA faculty at Hooke Park between 2016 and 2021.

6 Tree Fork Truss project credits. Students: Mohaimeen Islam, Sahil Shah, Swetha Vegesana, Vivian Yang, Zachary Mollica. Tutors: Martin Self, Charley Brentnall, Toby Burgess. Site Manager: Jack Draper. Robotic Developer: Pradeep Devadass. Hooke Park staff: Charlie Corrywright, Christopher Sadd, Edward Coe. Structural engineering: Arup.

7 Together, a physical measurement of the distance from camera to tree, the angle of inclination of the photo above the horizon, and the phone camera's field of view provided sufficient information to accurately correct photographs taken from ground level.

8 A deprecated Autodesk product, 123D Catch, was used to 3D scan forks in the Tree Fork Truss project. This tool is like many now available for smart phones in that it functions as a 'black box', with photos uploaded to the cloud and processed without manual control.

9 A detailed discussion of the workflow that enabled the Tree Fork Truss' use of non-linear elements is available in 'Tree Fork Truss: Geometric Strategies for Exploiting Inherent Material Form', a paper published with Design + Make founding director Martin Self in the proceedings of the 2016 Advances in Architectural Geometry conference.

10 Peter Wohlleben's 'The Hidden Life of Trees' and Suzanne Simard's 'Finding the Mother Tree' are two among many incredible books published on trees and forests in recent years. Each confronts us with the rather magical characteristics of these living beings we rely so much and often on.

11 As LiDAR tools have become accessible, their use in forestry has become more common. Uses of the technology tend however to directly derive from longstanding methods of conventional forest mensuration: with analysis primarily concerned with forest productivity measured in useable wood grown per acre.

12 Powers, 2018, page xx.

13 Stud-A project credits – Students: Georgina Bowman, Jacopo Silvestri, Sirisha Munnangi, Weikun Xu. Tutors: Zachary Mollica, Jean-Nicolas Dackiw, Jack Draper, Martin Self. Hooke Park staff: Charlie Corrywright, Christopher Sadd, Edward Coe, Jack Cardno. Structural engineering: Arup.

Bibliography

Hooke Park College, Dorset – House and Training Centre in Green Roundwood. In *Patterns 8*, published by Buro Happold.

Moe, Kiel. Think Like Forest: Maximising the Environmental Impact and Energetics of Building Timber. In *Rethinking Wood: Future Dimensions of Timber Assembly*, Eds. Markus Hudert and Sven Pfieffer. Birkhäuser, 2019.

Mollica, Zachary and Self, Martin. Tree Fork Truss: Geometric Strategies for Exploiting Inherent Material Form. Published in the *Proceedings of Advances in Architectural Geometry* 2016.

Powers, Richard. *The Overstory*. W.W. Norton & Company, 2018.

Ramage et al. The wood from trees: The use of timber in construction. Elsevier, 2017. In *Renewable and Sustainable Energy Reviews*. Vol 68, Part 1, P.333–359.

Ecotectonics

Using Digital Fabrication Toward Multispecies Design

Ehsan Baharlou

Introduction

The building and construction industries consume 40% of global energy and contribute 40% of global carbon emissions.[1] These percentages will only increase as we build more living spaces to accommodate population booms that are projected to occur until 2050.[2,3,4] Architects and builders can reduce the global energy consumption of a building's life cycle by not only modifying how we extract raw materials but also how we transport them to construction sites. However, what is most critical is that we find renewable material resources. If we fail to address the numerous environmental shortcomings of the construction industry, we will witness a scale of carbon emissions that we will not be able to overcome.

In recent years, novel building approaches such as circular construction and cradle-to-cradle have shown that they have the potential to move the construction industry toward zero carbon emissions. These approaches generally encourage rethinking the types of materials we build with, for example, repurposing waste materials such as agricultural byproducts. My own approach to the development of eco-friendly construction methodologies begins with using ecologically sound materials that are locally available. The methodologies I seek to develop can ideally be carbon-negative solutions that not only sequester carbon but also remove it from the atmosphere.

My research mainly focuses on using building materials that can readily have symbiotic relations with their ecosystems. I believe that architects should use materials

DOI: 10.4324/9781003296522-12

that are inherently ecological rather than attempt to impose ecological sensitivity on materials that have none. One ecological material that I believe has great building potential is soil. I am convinced that we can use emergent fabrication technologies to revolutionize how we use soil in construction, along with the environmental benefits of doing so. Further, and perhaps more importantly, I believe that advancing soil construction can revolutionize the construction industry itself.

Soil is a ubiquitous material that generally consists of organic matter, microorganisms, minerals, gas, and water.[5] Soil's ubiquity and constructability have allowed diverse species—from ants to birds to humans—to use the material to build their habitats for thousands of years.[6] Above all, soil has the capacity to nurture diverse ecosystems, such as those centered on fungi and plants.

To preserve the health of a soil ecosystem, and simultaneously adapt it to architectural applications, designers need to cooperate with other disciplines, such as environmental science and ecology. Such interdisciplinary collaboration can help architects begin to integrate ecological concerns into their design processes. In doing so, architects can advance beyond their typical collaborations with engineers and builders and move from human-centered design to "multi-species design."[7]

For instance, in my own digital fabrication research, I work with an interdisciplinary team in order to approach soil from what I call an "ecotectonic" standpoint.[8] In this approach, my team and I experiment with a specific mixture of soil and water in the creation of digitally fabricated structures. I believe we can couple soil's unique viscoelastic, muddy material properties with the precision of digital fabrication to open new possibilities for sustainable construction. Such coupling of ecological messiness with digital efficiency underlies the principles of ecotectonics.

Soil Tectonic Structures

A Brief History of Earthen Construction

Earthen buildings are typically found in arid climates around the world in which natural building materials like wood are less abundant. Shortages of both natural and imported material supplies prompt local communities living in these climates to be more resourceful with their building materials, for example, by looking to the very material beneath their feet: earth. Earthen buildings have traditionally been developed in one of several ways, including rammed earth, adobe blocks, and compressed blocks.[9]

Earthen construction has brought about certain cultural and social effects in the communities that employ its methods. These effects are directly related to the various advantages and disadvantages of earthen structures. For example, as mud changes from a viscoelastic to a solid state, it begins to demonstrate particular physical weaknesses. One of these vulnerabilities is the inability to withstand inclement weather, especially rainfall. When mud structures absorb water, they become bulky and heavy. This physical deformation eventually leads to breakage of the mud's inter-particle bindings. Another weakness of solid mud is its mechanical properties, especially its low tensile strength.

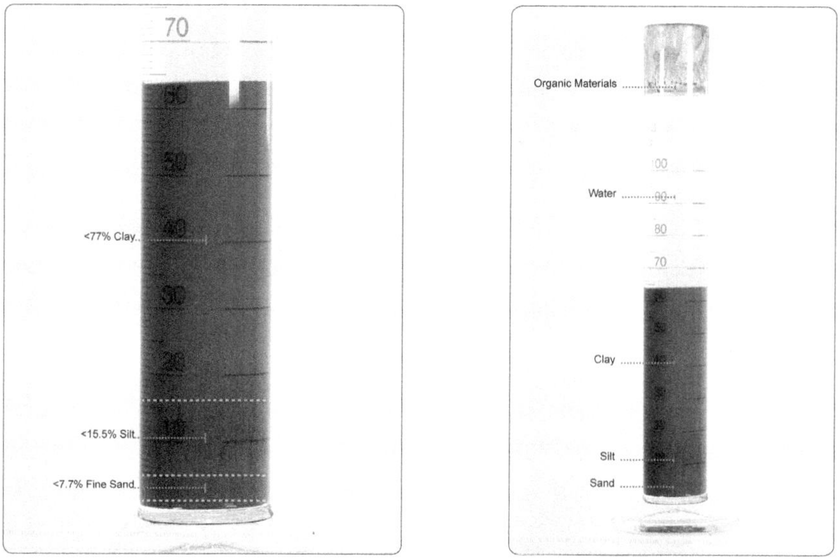

Figure 10.1 Sedimentation test to analyze the texture of the soil sample.

In traditional cob construction, builders improve the tensile strength of mud mortar by adding natural substrates like straw.

The constant repair and maintenance required by earthen structures demands intensive care and labor. Such incessant repair work has led to several innovative techniques for overcoming mud's weaknesses. For instance, analytical methods, such as sedimentation tests, enable builders to measure the soil's mineral characteristics before it is used in construction (Figure 10.1). Soil characterization determines the soil texture, which can allow builders to predict the overall behavior of the material when it is placed in structures. With a greater understanding of soil characteristics, builders can select appropriate additives for improving their pasty mixtures. Improving soil texture is not limited to human builders; other species, including birds, use additives, such as biopolymers like resin or saliva, to improve the properties of the soil they build with.[10] Builders can also add stabilizer to soil to improve its ability to withstand erosion and tensile loads. For instance, chemicals or mineral stabilizers like cement, lime, and plaster are traditionally added to enhance soil's inter-particle binding and, thus, mitigate erosion.

My own approach to the preservation of soil structures is to keep soil ecosystems active with organic agents. Such preservation is a kind of self-maintenance in which microorganisms and plants are introduced to the soil in order to maintain it; the soil, in turn, provides an environment in which these species can grow. Similar self-maintenance methods have proven to enhance the stability of earthen structures. In landscape architecture, for instance, plant and fungi roots are used to prevent soil erosion in berm construction.

In my digital fabrication research, I have explored the use of a fungi-based erosion prevention method in the 3D printing of earthen structures. My hypothesis was that the sprawling root networks of fungi and mycelium would be great at maintaining 3D-printed earthen structures. More broadly, I wanted to use this exploration to demonstrate the ways in which a multispecies design approach can be integrated with digital fabrication to evolve architecture. I believe that if we fully embrace earthy materials, we can begin to develop a basic ecotectonic model for the digital fabrication of earth structures. In other words, ecotectonics can emerge from the careful integration of soil cultivation and building construction processes. These tectonics can generate unique morphologies, similar examples of which can be seen in vernacular architectures in traditional earthen building in both Iran and Morocco. Such examples of vernacular architecture demonstrate the benefits of integrating seemingly opposite, or unrelated, material and construction processes in order to mitigate environmental factors.

Integrating Formation and Materialization

Traditional earth construction techniques are generally labor-intensive. In cob construction, for example, a complex arrangement of mud and straw layers requires multiple fabrication processes, all of which are normally performed without the aid of a machine. The application of digital technologies to such techniques can reduce labor and open space for innovation. However, in order to translate these traditional techniques to mechanized workflows, architects need to first gain a deeper overall understanding of analog earthen fabrication processes.

There is precedent for the industrialization of earthen construction; examples include the mass production of adobe bricks, the prefabrication of rammed earth walls, and the extrusion of wet loam.[11] My own research is broadly inspired by the techniques of traditional wet loam construction. In such techniques, the formability of the loam helps craftspeople manually stack layers of the material on top of each other and experiment with various tectonics—the process is akin to manual 3D printing. Wet loam can also be integrated into wood frame construction to produce building envelopes with unique structural and insulative features.

To encapsulate the human intuition required by traditional earthen construction, we need versatile machines, such as industrial robotic systems, that are capable of embracing the varying morphology of earth materials. The customizability of industrial robotic arms can enable the construction of multifunctional earth structures, for example, building envelopes composed of varied materials that have different environmental performances. A robotic arm equipped with an extruder can deposit different earthy materials at precise locations within a structure. In working with earthy materials, I have noticed that the nonuniform behaviors of earthy materials in robotic additive manufacturing challenge the zero-tolerance expectation of typical robotic construction processes.

The modulation of various parameters inherent to robotic 3D printing soil processes can influence both the aesthetics and structural performance of 3D-printed earthen structures. For example, by changing the definition of 3D print layers, we can augment the texture of 3D-printed soil structures. However, the distinct signature of 3D printers, those that print soil or some other material, is often revealed in their failures. For example, wavy surface patterns typically arise from a machine's vibrations: each wave is a result of one complete revolution of its motor. In my opinion, such texture is the articulation of the machine's agency. This agency is often expected to be eliminated from

Figure 10.2 The wavy pattern of the digital fabrication machine's "signature."

typical, industrially focused 3D printing processes, which aim to predefine the 3D-printed thing, from design to fabrication. However, there are always unintended tectonics, such as the wavy texture, which emphasize the inevitable interaction between matter and machine (Figure 10.2). We should learn to embrace such inevitable interactions in order to develop ecotectonic methodologies.

On the other hand, we can also work beyond the 3D printer's inherent parameters and use custom extruder tools to explore various tectonic structures. Such sophisticated tools can help us control the deposition of diverse earthen materials and, thus, can facilitate multi-material 3D printing. For instance, such a tool can combine natural and artificial materials in the formation of eco-composite façade systems. These ecotectonic structures can enhance the building envelopes by creating more selective transitions between interior and exterior environments: certain exterior elements can be admitted to the interior space while others can be repelled.

Ecologically Active Soil Tectonics

The development of eco-friendly building envelopes requires a tectonic approach that can simultaneously support ecosystem growth and address the needs of human inhabitants (Figure 10.3). The ecosystems that can be engineered into these composite structures can help us to sequester carbon and, therefore, move us, architects, from designing a human-centric ecosystem to a more inclusive, non-species centric one. This transition can result in the preservation of biological species that are useful for anthropocentric environments. Thus, these composite envelopes can support surrounding ecosystems while sustaining a living environment for both flora and fauna. Further, if we want ecologically active composites to perform in multiple ways, then we need the fabrication processes that produce them to be tuned to various material properties and requirements. To rethink soil

Figure 10.3 The timelapse of the germination on the earthen wall section after 48, 96, and 144 hours.

as a 3D printing material with an active ecology, we must address potential issues with soil's printability and structural capacity. We also need to develop a soil mixture that can provide a growth environment for biocompatible species that can serve human needs.

Similar to green wall façade systems, ecotectonic structures can help to filter pollution. Many types of green façades consist of separated vertical layers that carry out multiple functions, such as supporting structures, vegetation, irrigation, and drainage.[12] The synthesis of these diverse functions into one system must include the consideration of ecological, material, and fabrication parameters in design processes. Such integration emphasizes the significance of material organization in activating ecosystems in building envelopes. However, material agency can only be realized through an understanding of construction limitations. Thus, in 3D printing soil for example, the constraints of the digital tool must be taken into consideration as well.

The development of living tectonics with soil-based materials can, and should, be adopted as a form of circular building construction. Since the basic mixture of these structures consists of soil, sand, silt, and water, it is easily reusable. Soil materials that are not contaminated with non-reusable additives can be repurposed through crushing, sieving, and dry-to-wet mixing (Figure 10.4). Biopolymers with high biodegradation levels is an alternative to the harmful chemical additives that are used in conventional earthen construction. Biopolymers increase soil mechanical and rheological properties and reduce decay (Figure 10.2). Soil mortar properties should always be improved with some consideration of how plants may grow in the soil once it has been repurposed or recycled in the future.

In my research, I have demonstrated that 3D-printed earthen structures can maintain both their structural integrity and their fertility when biopolymers are mixed into the soil mortar (Figures 10.5 and 10.6). The ratio of soil to biopolymer must be calibrated with the amount of acidity and salinity required for the plants. For example, plant species like White Clovers have specific biochemical requirements for sprouting and seedlings. The plant growth rate exhibited by 3D-printed earthen structures is a strong indication of

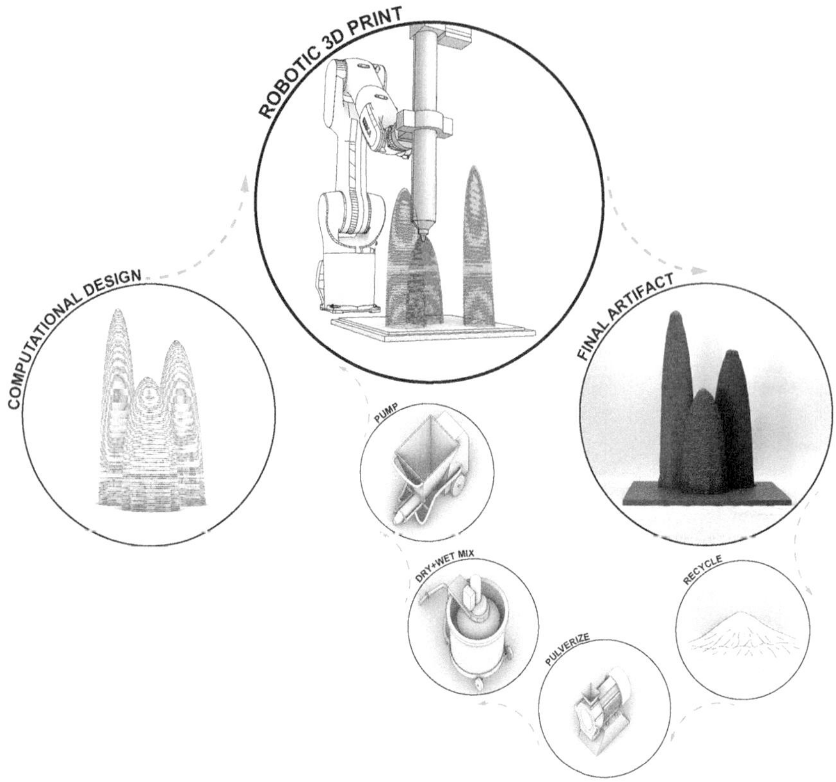

Figure 10.4 A generative diagram of robotic 3D printing—from design to fabrication—in a circular construction methodology.

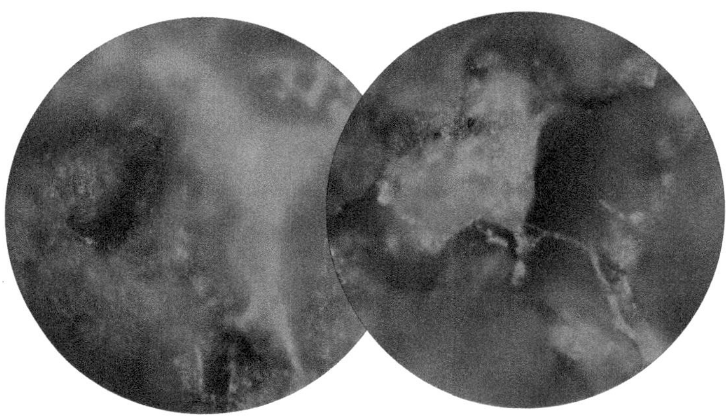

Figure 10.5 Microscopic images of the soil mixture without a binding agent (left) and with a biopolymer binding agent (right).

Figure 10.6 Three dome-like objects made from three different soil mixtures.

Figure 10.7 Detail of a printed dome made from a soil mixture that was improved by a biopolymer binding agent.

the level of ecological activity within their mortars. When the mortar meets three necessary conditions—maintaining structural integrity, being readily printable, and providing nutrients for vegetation—the 3D-printed earthen structure can be ecologically active and plant growth can thrive (Figure 10.7).

One architectural typology that is challenging to 3D print with soil, but is potentially good at fostering plant growth, is a dome (Figure 10.8). Because the semi-dry printed material must keep its water content until the seeds start to sprout, the ratio of seeds should be proportional to the dome's surface area, which can be mixed with the mortar or sprinkled in between print layers.[13] Successful seed germination on dome-like surfaces can lead to the implementation of 3D printing ecologically-active structures on larger scales (Figure 10.9).

The scaling up of 3D-printed soil envelope prototypes presents several challenges due to the material's inherent characteristics. In my research, I monitor the physical behaviors of 3D-printed soil layers as they are being printed in order to find the most critical fabrication parameters. One of these parameters is the water content of the

Figure 10.8 Robotic 3D printing of a dome with seeds.

Figure 10.9 Explorations into varying the height of the 3D-printed dome in order to expand the potential for seed germination.

mortar. Evaporation of water on the surface of 3D-printed soil layers creates a discrepancy between the moisture content of the soil and that of the surrounding environment. This discrepancy causes the 3D-printed layers to shrink on one side. In small-scale testing, such physical changes are negligible. However, in large-scale testing, the shrinkage becomes visually apparent and can ultimately lead to the breakdown of a structure.

The shrinkage issue can be overcome by modulating the ratio of the biopolymer in the mix. Larger 3D-printed soil prototypes potentially contain many ecological benefits, for example, by increasing the widths of printing layers, we can influence and improve the direction of plant growth. It is therefore critical that we continue to push the scalar limits of current 3D printing soil methodologies.

Toward Ecotectonic Architecture

Ecotectonics is a novel building approach that addresses the environmental challenges brought on by the construction industry, most notably climate change. At its most basic level, ecotectonics aims to reduce carbon emissions. However, this interdisciplinary approach also proposes the integration of different environmental performances within a building envelope, from greenery to insulation, all while maintaining the envelope's structural integrity. For instance, an ecotectonic wall can have a green cover on its exterior (Figure 10.10) and a suitable texture for architectural programs within an interior environment (Figure 10.11). The empty space between the inner and outer layers of such a wall can be activated with natural insulation materials, such as live organisms like fungi hypha (Figure 10.12). Further, the addition of a myco-composite to that same ecotectonic building envelope can enable us to explore the potential symbiotic relationships between distinct species, including humans and the flora and fauna that we are already surrounded by. These kind of ecotectonic construction systems can thus improve architectural performance in design criteria such as heat island mitigation, acoustical and thermal insulation, and even aesthetics.

Figure 10.10 The exterior face of a 3D-printed soil cavity wall is covered in germinated seeds.

Figure 10.11 The interior face of a 3D-printed soil cavity wall shows how a controlled material flow produces a repetitive texture.

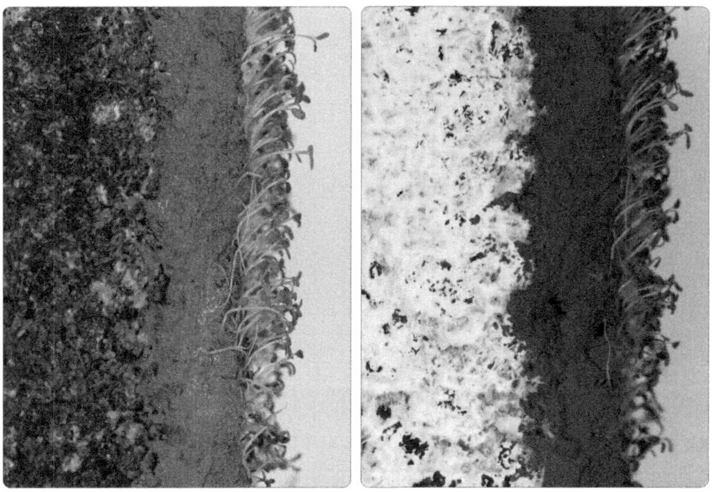

Figure 10.12 A section of a 3D-printed soil prototype that shows the symbiotic relationship between the growth of mycelium infill and embedded seeds after 7 days (left) and 14 days (right).

Notes

1 United Nations Environment Programme, "2021 Global Status Report for Buildings and Construction: Towards a Zero-emission, Efficient and Resilient Buildings and Construction Sector," 2021. Accessed: May 3 2023. [Online]. Available: https://www.unep.org/

2 F. Barbosa et al., *Reinventing construction: A route to higher productivity*. McKinsey Global Institute, 2017.

3 J. Woetzel, J. Mischke, S. Peloquin, and D. Weisfield, *A Tool Kit to Close California's Housing Gap: 3.5 Million Homes by 2025*. McKinsey Global Institute, 2016.

4 J. Woetzel, R. Sangeeth, M. Jan, G. Nicklas, and S. Shirish, *A blueprint for addressing the global affordable housing challenge*. McKinsey Global Institute, 2014.

5 P. M. Huang, Y. Li, and M. E. Sumner, Eds., *Handbook of soil sciences*. Boca Raton, FL; London; New York: CRC Press, 2012.

6 M. H. Hansell, *Built by animals: The natural history of animal architecture*. Oxford: Oxford University Press, 2007.

7 W. W. Weisser et al., "Creating ecologically sound buildings by integrating ecology, architecture and computational design," *People and Nature*, vol. 5, no. 1, pp. 4–20, 2023, doi: 10.1002/pan3.10411.

8 Ecologically active structures is an ongoing project in collaboration with Prof. Ji Ma (Department of Materials Science & Engineering at the University of Virginia) and Prof. David Carr (Department of Environmental Sciences at the University of Virginia).

9 United Nations Centre for Human Settlements (Habitat), *Earth construction technology: Manual on production of rammed earth, adobe and compressed soil blocks*. Nairobi: United Nations Centre for Human Settlements (Habitat), 1986.

10 Y. Jung, S. Jung, S.-I. Lee, W. Kim, and H.-Y. Kim, "Avian mud nest architecture by self-secreted saliva," *Proceedings of the National Academy of Sciences of the United States of America*, vol. 118, no. 3, 2021, doi: 10.1073/pnas.2018509118.

11 G. Minke, *Earth construction handbook: The building material earth in modern architecture/Gernot Minke*. Southampton: WIT Press, 2000.

12 M. Manso and J. Castro-Gomes, "Green wall systems: A review of their characteristics," *Renewable and Sustainable Energy Reviews*, vol. 41, pp. 863–871, 2015, doi: 10.1016/j.rser.2014.07.203.

13 S. Barnes, L. Kirssin, E. Needham, E. Baharlou, D. E. Carr, and J. Ma, "3D printing of ecologically active soil structures," *Additive Manufacturing*, vol. 52, p. 102670, 2022, doi: 10.1016/j.addma.2022.102670.

Bibliography

F. Barbosa et al., *Reinventing construction: A route to higher productivity*. McKinsey Global Institute, 2017.

G. Minke, *Earth construction handbook: The building material earth in modern architecture/Gernot Minke*. Southampton: WIT Press, 2000.

J. Woetzel, J. Mischke, S. Peloquin, and D. Weisfield, *A Tool Kit to Close California's Housing Gap: 3.5 Million Homes by 2025*. McKinsey Global Institute, 2016.

J. Woetzel, R. Sangeeth, M. Jan, G. Nicklas, and S. Shirish, *A blueprint for addressing the global affordable housing challenge*. McKinsey Global Institute, 2014.

M. Manso and J. Castro-Gomes, "Green wall systems: A review of their characteristics," *Renewable and Sustainable Energy Reviews*, vol. 41, pp. 863–871, 2015, doi: 10.1016/j.rser.2014.07.203.

M. H. Hansell, *Built by animals: The natural history of animal architecture*. Oxford: Oxford University Press, 2007.

P. M. Huang, Y. Li, and M. E. Sumner, Eds., *Handbook of soil sciences*. Boca Raton, FL; London; New York: CRC Press, 2012.

S. Barnes, L. Kirssin, E. Needham, E. Baharlou, D. E. Carr, and J. Ma, "3D printing of ecologically active soil structures," *Additive Manufacturing*, vol. 52, p. 102670, 2022, doi: 10.1016/j.addma.2022.102670.

United Nations Centre for Human Settlements (Habitat), *Earth construction technology: Manual on production of rammed earth, adobe and compressed soil blocks*. Nairobi: United Nations Centre for Human Settlements (Habitat), 1986.

United Nations Environment Programme, "2021 Global Status Report for Buildings and Construction: Towards a Zero-emission, Efficient and Resilient Buildings and Construction Sector," 2021. Accessed: May 3 2023. [Online]. Available: https://www.unep.org/

W. W. Weisser et al., "Creating ecologically sound buildings by integrating ecology, architecture and computational design," *People and Nature*, vol. 5, no. 1, pp. 4–20, 2023, doi: 10.1002/pan3.10411.

Y. Jung, S. Jung, S.-I. Lee, W. Kim, and H.-Y. Kim, "Avian mud nest architecture by self-secreted saliva," *Proceedings of the National Academy of Sciences of the United States of America*, vol. 118, no. 3, 2021, doi: 10.1073/pnas.2018509118.

Failing Matters

An Interview with Skylar Tibbits

ZACH COHEN: Let's start by talking about the work of your lab at MIT, the Self-Assembly Lab. To me, the Lab is actually somewhat of an outlier within digital fabrication in architecture because it more heavily relies on the intelligence of materials than the intelligence of technologies, or, to put it another way, I think the Lab treats materials *as* technologies. I'd be curious to know if you see your research approach in a similar way and, if so, what you see as the role of digital fabrication technologies in the Lab's work.

SKYLAR TIBBITS: I think there are three categories of work that we've done over the past decade: one is on programmable materials, one is on fabrication technologies that enable us to *make* those materials, and the third is on construction- or large-scale processes of self-assembly. In the beginning, we would use all sorts of fabrication technologies, whether those were milling machines or laser cutters or robotic arms, and we used them just as tools to produce the materials, geometries, or behaviors that we wanted to produce; the idea behind 4D printing, for instance, was that we would use existing machines (3D printers) to produce new materials. In that way, digital fabrication was not really the research, it was a means of production to achieve material behaviors. But then we started to realize that the existing machines weren't giving us what we needed. I think part of that was that there's this spectrum: from very high-end, expensive machines that are hard to use as research tools and really meant for production or prototyping to low-end machines that are great

 DOI: 10.4324/9781003296522-13

for hacking and playing and experimenting, but are poor quality. And so that's when we decided we need to develop our own processes so that we could experiment, play, change materials, change scale, and have all the precision and great qualities of digital fabrication machines, but also have total freedom. That's where projects like rapid liquid printing came in and, eventually, liquid metal printing. Now a third of our research is just developing these new kinds of fabrication processes. And that's because we still want to create new materials, but also because we want to push the field forward. I think our collective ideas of additive manufacturing or fabrication, even, often get constrained by what machines are out there. Instead, I think we should make our own nuanced, different machines that can do things we've never seen before.

ZC: Would you then say that the work has pivoted, or maybe evolved, from appropriating whatever fabrication technologies were readily available to making fabrication technologies of your own?

ST: I think it evolved because we needed new machines and new capabilities, and then started to realize that there were interesting opportunities to make these new technologies and processes that weren't available, and then that, in turn, led to new projects on materials.

ZC: I'd be interested to hear if and how the idea of "material intelligence" has also evolved in your work, because it seems like there's a delicate balance that you've had to strike between developing specific tools to handle or control materials and letting those materials emerge on their own. How would you say you've found that balance? And has that balance changed over time (Figure 11.1)?

Figure 11.1 The Self-Assembly Lab's research into granular jamming uses little more than rocks and strings to create high-performance architectural structures. © Self-Assembly Lab, MIT.

ST: There are a lot of different ways to look at that. For example, one way would be to say the material has total agency and freedom to do whatever it wants—we can call that the "out-of-control approach." And with that approach, you can make weird, wild things. On the other end of the spectrum there's the approach in which you say: "we want the material to do this or that, and I'm going to tell the material to do that, and please, material, show me that you can do it." But something in the middle of those approaches is the collaboration with materials that we're interested in. In that more balanced approach, we give materials some information, goals, or constraints, and then let the materials have agency. And through that combination, something emerges that is precise and repeatable, but also surprising and different than what we would have necessarily prescribed. I think we're most interested in that middle, but, sure, sometimes we like things that are random, crazy, and out of control and other times things that are very prescribed.

In the beginning, we used pretty specific materials that required specific fabrication techniques, and only through those particular combinations could we achieve new material behaviors that could sense or respond. Over time, I think our approach has gotten more and more simple in a way, like with the jamming projects that just use rocks and string or our "Growing Islands" project in the Maldives that just uses sand and water. But another path has also emerged, like in our knitting work, that has gotten more and more technical, for better or worse, like developing fibers from scratch that have just the right properties, cross sections, and colors in order to have the right behaviors and right activation ranges. So the work is going in both directions (Figure 11.2)!

ZC: That's interesting because it seems like both worlds—the more entropic world of piling rocks and string and the more controlled world of determining individual stitches—play out at different scales. For instance, you can't necessarily work at the

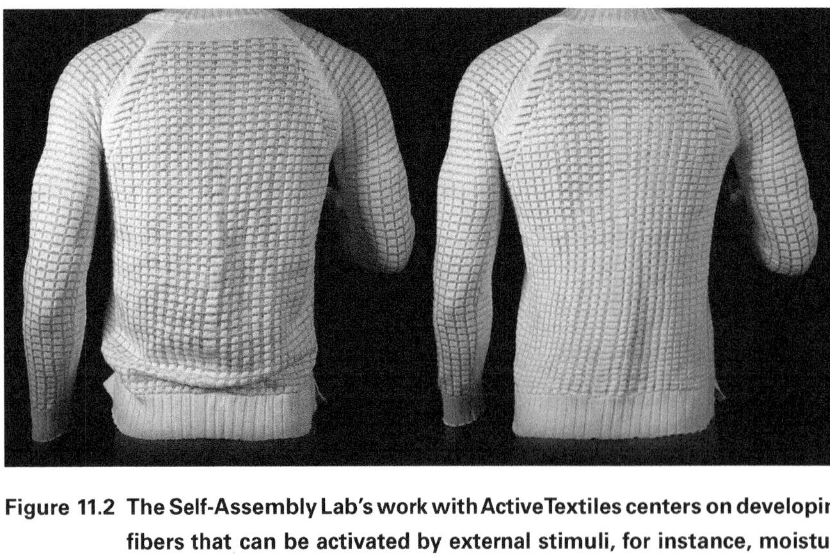

Figure 11.2 The Self-Assembly Lab's work with Active Textiles centers on developing fibers that can be activated by external stimuli, for instance, moisture or heat, to create clothing that can mediate between people and their environments. © Self-Assembly Lab, MIT.

Figure 11.3 The vast oceanic site of the Lab's Growing Islands project in the Maldives required the use of locally sourced, readily available materials that already exist in bulk. © Self-Assembly Lab, MIT.

scale of an individual stitch when you're trying to form an entire dunescape. But do you think there's another world in which you start to manipulate bits and atoms to work at a middle scale, the scale of architecture perhaps? To put the question in a different context, there's a whole body of research looking at how to make "smart environments" by embedding microcontrollers and sensors into spaces that we occupy. Do you think that's a viable approach to working small but having an architecture-sized impact (Figure 11.3)?

ST: I think one of the reasons that our rocks and string projects or our Maldives project developed with these very simple material palettes is purely because of scalability. Early on in those projects, we would try different combinations of lots of weird materials. We even had a whole section of research dedicated to smart components, which were not at all "smart" in the sense of sensor technology, but more *geometrically* smart. And then we realized that's not really scalable: you can't build large-scale structures that need millions of custom parts that need to be digitally fabricated. Working that way really limits what you can do because there's a lot of waste, cost, production time, etc. The way we start projects now is often by asking ourselves, what materials can we get in large quantities that are widely available and in abundant supply? That way we're not just producing truckloads of waste. And, so, scalability is the reason we started looking at those more simple, fundamental materials.

I don't know if you can't have a larger impact with digital fabrication technologies. For instance, maybe there's a way to create a big impact by being more strategic, and not necessarily needing more quantities of material or bigger technologies. But in our projects, I think it's more that we develop materials and activation

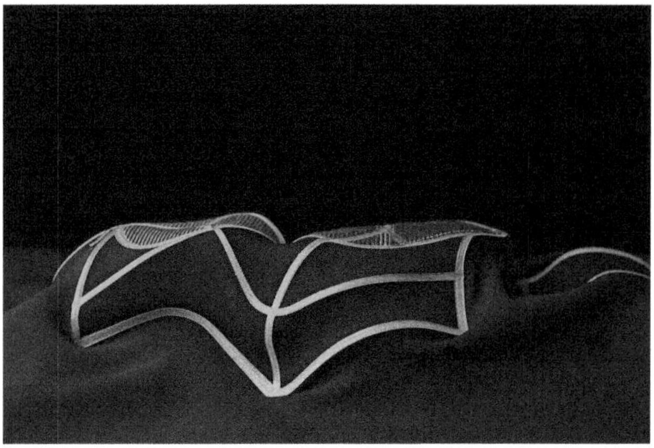

Figure 11.4 In the "jumping textiles" project, the Lab developed a process for creating shoes through 4D printing. © Self-Assembly Lab, MIT.

energies tailored to the scale we're working at or the application we're working on. We'll ask ourselves: what is this thing that were working on? What does it need or want to be? What materials are available? And what energy is available in those materials?

ZC: That "what is it" question points to something else that I'd be interested to hear your take on, which is the role of human intuition in cultivating material intelligence. At what point, and to what extent, does human intelligence or intuition play in the development of material intelligence? In other words, how frequently are the "what is it" or "what can it be" questions asked in the research and at what moments (Figure 11.4)?

ST: If the question is "what is the application" or, even, "what is the final scale," that's often given. We'll know, for example, if we're going to be working on a shoe or a seat in a car or trying to make an island. How does it look? How does it behave? How does it perform? And how does it become what it needs to be? Those are the questions we're interested in. And that's mostly because of collaborations with industry partners, who say the goal is *this*. We don't know how to get to *that*, and that's what we work through.

Sometimes it's not the case, for instance, we had this one project in which we were testing the self-assembly of floating particles in an air chamber. There, we were most interested in exploring the air chamber as a kind of lottery tumbler and seeing what assemblies would come out. What we found is that not only did different structures start to emerge, but they also began to hover and fly. And, so, that was an "aha" moment because flying *became* the application: we inadvertently developed structures that learned to fly.

A lot of our early 4D printing research was also like that, and so was the "jumping textiles" project we did with the designer Christoph Guberan. In that project, we didn't start with a shoe, we started with the process and then tested a bunch of different geometries in order to build an intuition of what the process could produce. In the middle of this testing, we were asked to contribute to an exhibition on the future of the shoe and thought that was perfect because this jumping textile process enabled us to easily create complex curvatures, which is one problem in typical shoe production.

Getting back to the question of intuition though, I think what our lab is about is trying to play and experiment and learn enough so that we can build an intuition of what's possible. Even with the Maldives project, we could have simulated the water dynamics or tried to test things out in the field as much as we wanted, or at least as much as time and money would allow. The goal, however, wasn't really to know exactly what would happen but rather to build an intuition of what *might* happen. So how do you build that intuition? That's where we use just lots and lots of pure trial-and-error testing: test after test after test...and then through that testing, and by questioning why certain tests work and others don't, we can start to cultivate an intuition about what adjustments to the fabrication process may be productive.

Of course some people that come at this kind of research from more of an engineering standpoint might ask us: "why would you do that? Just simulate it until you understand everything about it and then fabricate it." Or other people might say "Trial and error? That's not research." But to me, that's everything. That's how you build the intuition: you just have to put the time in and build and play and test and learn.

ZC: That brings up what I see to be another unique aspect of the Lab's approach, which is that you're open about the role of failure in your research. Failure is often treated as a somewhat taboo topic in the world of digital fabrication or, really, in any kind of scientific or pseudo-scientific discipline. But for you, failure appears to be something that is necessarily embraced and learned from, and that's rare! For example, you wouldn't go to a digital fabrication conference and see everyone present on what *didn't* work.

ST: That doesn't happen.

Figure 11.5 Testing structures to failure, and learning from those failures, is integral part of the Lab's research process. © Self-Assembly Lab, MIT.

ZC: Yes, it doesn't happen! And there are probably some pretty obvious reasons why it doesn't happen, like funding and industry partnerships, but I'd be curious to hear your take on why exactly failure has been so shunned from the conversation, and whether or not you think its exclusion is hindering the true progress of digital fabrication as a subdiscipline in architecture (Figure 11.5)?

ST: It's a really interesting question, and one that I've had so many thoughts on—it's been a constant struggle. I think research is ultimately based on failure, it's fundamental to what we do. And every time we fail, we learn a lot more than when we succeed. So, we are always failing and building and playing and testing. And a lot of people want to talk about celebrating failure, like even here at MIT Architecture, we made a book called "Testing to Failure," which was meant to represent our guiding ethos. But I'm skeptical that people truly believe that. In fact, I actually often see the opposite of that. For instance, you can't be invited to exhibit at the MOMA or the Venice Biennale, and then show up with a failure. That doesn't work. No one accepts failures, even when the curators that put together those kinds of exhibits talk about celebrating failure, and even when we have books about it. People don't want to see failures and they don't want to show failures; they want to show success. that's the reality of it. Or maybe it's that if you show successes, you can also show failures, but only as little failed prototypes, and only on the side, and only if it you call it "process." But that basically means that you can't take much of a risk, or enough of a risk, because you're expected to succeed. How much risk can there be when you know you're going to succeed?

We've failed, sometimes spectacularly, and lots of times just daily. In those larger failures is where you begin to realize that the system of exhibitions, deliverables, industry, collaborators, grants, funders, etc. are not set up to truly accept failure, no matter how much we may have said otherwise.

You can test and fail behind closed doors and, eventually, when you have successes, you can curate them and put them out in the world, and those are the things that turn into exhibitions, installations, buildings, landscapes, companies, etc. But you better not release the ones that aren't going to succeed!

ZC: That's my hunch as to what happens in most other digital fabrication research, but of course we don't see any of the curation process because the only projects that are released to the world are successes and even those projects are heavily redacted, or presented in such a careful way that any implicit failure is covered up.

ST: I think most of the time you need to read between the lines. Something that most of us in the discipline, including our lab, is guilty of is trying to document the project in the most beautiful way and tell the right story. Of course, there are always things that aren't successful or didn't go as planned and, as we said earlier, we often don't celebrate those. Whenever you see digital fabrication projects get published, you also have to ask yourself, "how much am I not seeing?" It's tough though, because who wants a video of something that is *just* okay or something that just *casually* fails. The work wouldn't be as exciting.

ZC: To me, the irony is that this collective allergy to failure may actually be what's preventing digital fabrication from leaving architectural academia and becoming more than just research. We make all these exciting things but when it comes time to be honest about the potential implications or applications of them, we often have to backtrack and say, "well, it doesn't *actually* do that" or "it's not *quite* ready for use." In other words, digital fabrication looks really good when it's incubated within academic institutions but once it leaves that bubble, it can't actually survive because we haven't been honest about what it is or what it can actually do. Sometimes I wonder: is digital fabrication *just* a hype machine?

ST: It would be good to more properly define what we're talking about when we say "digital fabrication." If we forget about architecture, for a moment, we'll see that digital fabrication is ubiquitous in every form of manufacturing, from signs to shoes to cars to planes to any plastic part; digital fabrication is in every single one of the factories that make those things, whether it's robotic arms, laser cutters, or even 3D printing. I think there's no question that that use of digital fabrication is a success: in those settings, those machines do what they are engineered to do, and have for a while. But when we're talking about digital fabrication here, I think we're talking about fringe or newer forms of fabrication, whether it's rock jamming or fiber winding. We're talking about various architectural experiments that test new ways to fabricate and build things. In this kind of work, we're pushing the boundaries of what's possible. So even if the application isn't immediately clear, the goal in our world of digital fabrication is to be scalable in terms of literally pushing the physical scale rather than to be scalable in terms of manufacturing, efficiency, and production (Figure 11.6).

Figure 11.6 The Lab's work focuses on pushing the physical scale and application of experimental fabrication processes. © Self-Assembly Lab, MIT.

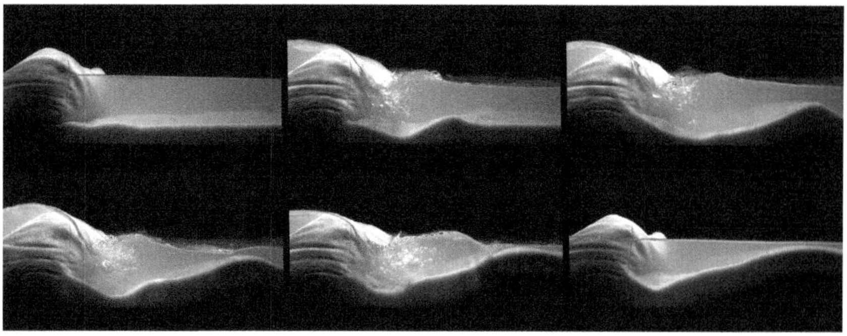

Figure 11.7 Early tests using sand and a wave tank helped to dream up the possibilities that would lead to the Growing Islands project. © Self-Assembly Lab, MIT.

ZC: I think that's really a good point, and part of what we're after in this book, which is that we should be more precise with what we mean by digital fabrication in architecture. I've recently started to use the term "architecture-oriented digital fabrication" because I want to make clear, as you said, that digital fabrication in our world means something different; it's also in a very different stage of development in architecture than it is in other industries or other disciplines that have already successfully adopted these processes.

ST: In other industries, it's sometimes helpful to talk about failure or experimentation, for instance, look at fashion. I think digital fabrication in architecture is like Paris Fashion Week in fashion design: we don't talk about producing clothes at scale, like Gap or H&M, or fast fashion or even sustainable fashion, like Patagonia. We're like Iris Van Herpen or Hussein Chalayan. And there's the good and bad of that kind of work: on the one hand, it's pushing the limits of what's possible, and on the other, it's not scalable, or realistic, or for the masses, or about production—it's about dreaming about new possibilities (Figure 11.7).

But then if you look at the automotive space, there's no room for failure. Automotive manufacturers are expected to come to auto shows with concept cars that work and that accurately show us what the future will look like. Consumers provide manufacturers with feedback on their concepts, and that ultimately leads them to the kind of innovations they are going to push through production—it's a process that's very grounded in what's needed and what works. And then eventually when cars are produced, they need to meet safety regulations that are as strict, if not stricter, than those in the building industry. Still, the automotive industry pushes forward in terms of technologies, fabrication, materials, performance, etc. So I'm not sure it's a lack of failure that's stopping our industry as much as it's the relationship we've built between innovation and production.

Sometimes I feel like digital fabrication is just cycling through ideas, and architecture is ultimately going off in two different directions: the innovative people go one way, and the rest of architecture goes the other.

ZC: I think another potential way to consider the value of digital fabrication in architectural discipline is from a conceptual standpoint. At the "Being Material" conference at MIT in 2017, you spoke about how the Digital can actually enable us to better understand the concept of material better, or even be material *better*. And so, now, particularly when so many people in the world are pushing deeper toward the Digital, or toward using the Digital to be more digital, for example those making Non-fungible Token (NFT) architecture, it makes me think that digital fabrication could be something that pushes back on that: in digital fabrication, the Digital could be used toward being more material.

ST: I think that is the keystone of the work, because for a long time there were these digital studios in architecture schools, like the "paperless studios," that used all the new computation and simulation software at the time, and we thought that that would be the future: a future in which we don't need material. (That was around the time Nicholas Negroponte's *Being Digital* book came out, which was the inspiration behind the *Being Material* conference.) But then, digital fabrication and Arduino and other technologies that bridge the gap between the Digital and the Material arrived on the scene, and there became a renewed interested in material. So I think the more digital we get, the more the pull of the Material is going to bring us back. It's not always the case, for instance with Augmented Reality/Virtual Reality (AR/VR) and all the people that just want to stay in their headsets all day. But those people are still in a physical environment and they still have material devices that they're wearing that need to be designed. There are generally all these other material parts of the Digital that we tend to not talk about. But for the most part, the Digital, i.e., computational tools and programming, led people to Arduinos and digital fabrication, which led them back into squirting, extruding, and melting physical materials. I see that as a good thing!

When we started the Lab, we asked ourselves, what is it about code and information that has this pull and how do you create a program not from software, but just from information and logic, and then how can you use that program to

Figure 11.8 Programming materials, as seen in the Active Textiles project, are central to the Lab's work and reflect the Lab's foundational belief that the Digital can be used to more directly connect to the Material. © Self-Assembly Lab, MIT.

directly connect to materials? So digital fabrication becomes the keystone or the connector between these two worlds—the logical and the material—and I think we should celebrate that (Figure 11.8).

ZC: What do you think about a relatively recent trend in which architects are not using fabrication to bridge between the Digital and the Material, but instead simply fabricating the Digital, for example, NFT houses or VR worlds? Is this just a momentary fad or do you think this movement has some staying power?

ST: I think probably a little of both. There are lot of people now into artificial intelligence and machine learning, AR/VR, or the NFT world that are staying in the Digital. Then there are a lot of other camps that prioritize the Material. I don't think one group will necessarily give way to another, but rather that we'll see lots of strange hybrids start to emerge.

I also think there are a lot of physical, material repercussions or constraints of AR/VR or the NFT world that we don't tend to think about, whether that's the servers, computing resources, energy, or the material devices that these digital technologies tend to require. I think we often think about these physical components as digital things because they're in service of the Digital, but they're physical! So maybe we can turn the dial up on our thinking about the relationship between the Physical and the Digital, especially in these worlds that are seeming purely digital.

ZC: It makes me think of something you said earlier in terms of how digital fabrication technologies expand the field of architecture. These newer technologies, like AR/VR, also appear to expand the field, or at least increase the amount of potential camps within it, but I sometimes wonder if what appears to be expansion is actually distraction. For example, returning to what you said earlier about the need to be more exacting in our definition of digital fabrication in architecture, I wonder if these relatively new movements are further distracting us from that collective definition? And then I wonder if we actually need that collective definition! Do we all need to get on the same page about we're talking about and moving toward, or should we simply appreciate the ever-expanding variety of definitions?

ST: I feel like historically there was the CAD tools camp and the BIM camp, and then there was the emergent technologies camp, followed by the digital fabrication camp, and then the robotics camp, and so on. Whatever the latest thing is, there's usually a camp of people interested in exploring it! And then eventually, that thing either fades out, becomes a focus of specialization, or becomes so ubiquitous that we take it for granted. For instance, everyone has laser cutters now and everyone uses CAD and BIM tools, so we don't need conference sessions on those technologies anymore—they've just become a part of the discipline. And I feel like it will always be like that: there will always be new strands that get folded back into the discipline, and then even newer strands that emerge from those folds, so I'm not sure we could ever focus every one or get on the same page. I think it's natural for

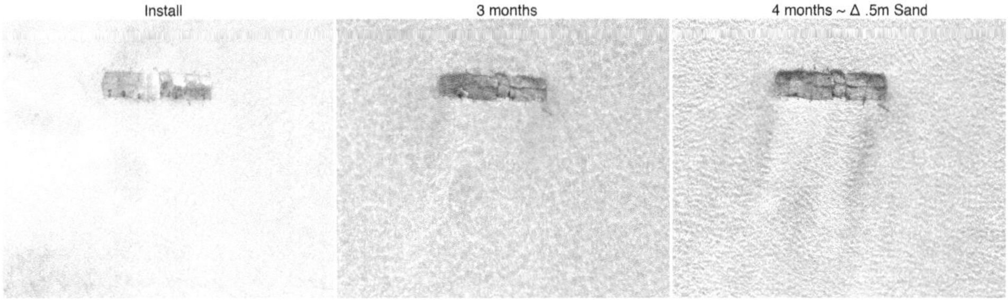

Figure 11.9 The initial test of the Growing Islands project demonstrates in which designers can collaborate with the forces of nature, e.g., wave currents, to invent processes for sustainabe construction. © Self-Assembly Lab, MIT.

these technologies to emerge and interests to come and go; they're all contributing to the larger movement of this self-organizing body we call architecture.

At the same time, there may be some benefit in having a more well-defined direction that we're headed in or goal that we're headed toward. For example, what if the entire community of computation and digital fabrication in architecture set out to tackle climate change and that became the North Star of the work. Then the NFT-makers could do their thing, the roboticists could do their things, and the materialists could do their thing, but it would always have to come back to this concrete goal that we collectively decided to move toward. That might be helpful. Rather than one camp doing their thing in isolation over there and another over there, and all aiming to solve different problems, we could all collectively move this ship in a certain direction.

ZC: That's a really interesting thought. To that point, what is orienting your own work these days and is it different from what your orientation was when you started the Lab (Figure 11.9)?

ST: It has definitely changed. But right now, I think our Maldives project is the closest thing to our North Star. For me, the questions of that project are really well aligned with the guiding questions of the Lab, for example, how do we truly collaborate with the forces of nature? How do we scale our work to have impact in global problems such as erosion and sea level rise?

Earlier we spoke about this radical and relevant spectrum that I'm always thinking about and I think that the Maldives project is the perfect combination of being both radical *and* relevant: it's radical, surprising, and beautiful in its approach, but it's also more important than any project that we've ever done because if we can really solve these issues, we can make a big difference.

There's been a lot of other projects we've done in the past that have been beautiful and interesting, but probably won't make a difference in the world in the long term. I don't think we can just make stuff for the sake of novelty anymore. We can't just keep making big plastic pavilions! If you want to do that, then tell me why it matters. Before it was enough to just be exciting, inspiring, different, or new, but

Figure 11.10 The Lab embraces the potential of material and fabrication research to be a form of "serious play". © Self-Assembly Lab, MIT.

now the work needs to be important in some way that's going to make a difference. I don't think it's just the case with digital fabrication, either, I think the discipline at large also struggles with this.

ZC: I know what you mean. I think, in digital fabrication at least, it begins with being self-critical and being candid about what these technologies actually can do, as we've been saying throughout this conversation. Otherwise, our collective fixation on novelty will continue to dominate and digital fabrication will remain this hype machine-type thing.

ST: I think with a little bit of distance and disciplinary critique, we can finally ask ourselves, does this work actually move the needle? Does anyone care about what we're doing? And that's where I think it would be helpful to align on a common direction.

As a counterpoint to those questions, I also sometimes think, does it all need to be so heavy? Can we just play with these technologies because they're enjoyable and fun? Maybe the self-criticism changes over time, and we're just in a moment of seriousness. There's also just value in doing fun stuff simply because it makes people happy? We shouldn't be so righteous about it!

Still, I feel like it would be useful if we had a collective goal. And maybe that goal just needs to shift every year and every year we need to take some time to update it (Figure 11.10).

A Robot Named Heidegger

Mark Jarzombek

One of the disasters of Western philosophy has been the hard separation between mind and body. Historians of philosophy mostly point to Plato as the culprit. Earlier Greek philosophers had a more complicated notion of human agency. Our livers, lungs, and even our gut (preserved in the colloquialism 'to have a gut feeling') were all elements of consciousness. Our internal organs were psychic entities that functioned upon the human and that even controlled certain types of behavior.[1] In the *Iliad*, Achilles on the verge of killing Agamemnon (which he eventually does not do) is described as being led by his heart *(etor)* that "within his shaggy chest went back and forth," which he sought to co-ordinate with his lungs *(phrenes)* and spirit *(thumos)*.[2] One can imagine Achilles – adrenaline pumping – in anxious anticipation of what he is about to do. *Phrenes* is associated with lungs in the sense of a warrior who needs to control his breathing before an attack.[3] It serves as a pilot to the *thumos*. *Thumos* is an internal urge or dynamic spiritedness, as we might mean with the words, a spirited horse. The word is also associated with the desire for recognition. In other words, Achilles in all of this did not possess a singular "body," but was a vessel for different almost quasi-independent forces that were not even really in alignment.

All of this was cleaned up by Plato and so it is not accidental that in the generation after Plato we already have Greek writers imagining the first automatons. One called Talos, as described in the *Argonautica* (3rd century BCE), was a bronze robotic soldier possessing emotions and judgments who was charged with protecting the homeland. Two thousand years later, we are still telling some version of that story, whether it be in *I-Robot*, *Terminator*, or *Transformers*.

DOI: 10.4324/9781003296522-14

The reason we have remained fixated on the humanoid-automaton project is that the mind-body dualism on which it is grounded was reinforced in the Middle Ages through neo-Platonic, Catholic philosophy and later by the conventions of the Enlightenment. René Descartes held that the mind was a nonphysical – and therefore, non-spatial – substance identified with self-awareness. Though there are a number of other philosophical arguments about the status of the mind, this dualism cemented itself into the general culture of Western world. It became natural – and easy – to think along those lines. The moto of MIT is *mens et manus*. This even extends into the early 19th-century distinction between organic and inorganic. The robotic body is on the side of inorganic, whereas intelligence, on the side of organic, is the locus of life. The Germans accordingly have two words for body, *Leib*, which corresponds to the living essence, one with "feelings," and *Körper* which is its mechanical aspect.

Artificial intelligence (A.I.) is now promising to complete the job of unification. With a better brain, the robot part of the equation becomes ever more serviceable. And indeed, robots now do astonishing things: laser-cut steel, manufacture cars, defuse bombs, and stack bricks. A recent headline in the New York Times stated "Who will take care of Italy's elderly? Robots, maybe."[4] In some regions of the world, they serve as sex partners. "The long wait may soon be over," says one advertiser. "Life-sized sex dolls combined with animatronics and artificial intelligence are starting to hit the market."[5] Data, the android who first appeared in 1994 in Star Trek, was the predecessor for this conversation. He was, as he points out, "fully functional" even in aspects of love.

The problem with the Platonic model has always been, however, its stubborn allegiance to the potential thoughtlessness of the body and thus to the machine. Regardless of how much A.I. is pumped into a robot, popular culture will never shake off the equation between robot and stupidity. Already Immanuel Kant was worried that a mechanical bird would deflect from our appreciation of the real thing. In a clichéd line from *I, Robot*, Will Smith said, "Get out my way, you dumb robot." And it is easy to find comments on Twitter like "I'm back at the grocery store and that dumb robot is trapping people who just want some creamer."[6] At best, as we hear from Zoe Smith, a character in *Lost in Space*: "The robot's job is to give the master what it wants."[7]

Post-Platonic European philosophy allowed us to fetishize the power of intelligence and belittle the power of the body or – as an extension of that proposition via the figure of Talos – to try to upgrade "the body" with intelligence, "feelings," and the simulacrum of meaning. *Westworld* put the two ends together, except that the robots here do not just possess "feelings" (the classic Post-Platonic, modernist problem) but have feelings that rise to the level of socio-political. Though the narrative plays on the standard cultural anxiety of robots-taking-over, it allows viewers to "side" with the robots as they struggle – like some futurist aboriginal culture – to find their own identity and special place in the dataverse.

So the question is – given that we are now in the 21st century and can begin to distance ourselves from the long Platonic nightmare – can we go back in time, so to speak, and imagine formations that avoid the teleology of dualism. Can there be an artificial Gut Intelligence – a computer that works not on 1s and 0s, but with acids and microbes? I think here of the *Shit Machine* (2000) of Wim Delvoye, a Belgian artist known for his inventive and often shocking and repulsive projects, in this case one that mimics the action of the human digestive system by converting food into feces in a public museum. Fungal computers are being produced at the Unconventional Computing Laboratory (founded in 2004 in Mexico) that aim to decipher the language fungi use to send signals through biological networks.[8]

Are there machines that make a mess out of the mind-body dualism? Survival Research Laboratories, an American performance art group founded in 1978 by Mark Pauline, had a show in 1982 in San Francisco – *A Cruel and Relentless Plot to Pervert the Flesh of Beasts to Unholy Uses* – that integrated machines with objects such as mummified and dissected animals and a robot that was part metallic dog, part cadaver.[9] I also think of Roxy Paine, a US sculptor whose Erosion Machine, 2005, consists of a robotic arm that traces and cuts patterns into large blocks of stone. The course of the arm's movement is determined by data sets, such as weather conditions and school test results.[10] What if robots self-destructed into great heaps of rubbish to transform earth into something like that fantastic, planet-of-garbage, Sakaar from *Thor: Ragnarok*?

The building industries pose a formidable challenge to the potential to transcend the cliches of Talos. The old Platonic line that distinguishes architecture from builder – that still defines so much of architecture to this day and that rises directly from the old mind-body dualism – mirrors itself in the difference between the world of Computer aided drafting (CAD) from the world of digital fabrication. CAD is, as we all know, far ahead of the game, but "intelligent" robots are becoming more standard. In Australia, the firm FBR made what it calls "a great big robot" that builds concrete block structural walls about ten times faster than humans. "It could change everything," according to the contractor, who added: "There is no difference [in look] between a human construction and a robot construction except that the robot construction is more accurate."[11] That is, of course, not true. The robot basically obliterated at least 30 people spread out of at least ten different trades.

The issue here is something else. From a philosophical point of view, the brick-laying robot is still a version of Talos, even if now cleverly civilianized. Is this really the end of the story, one that now celebrates the bluetooth world of humans and hydraulic machines that do our "labor"?

To avoid the compulsion to perfect Talos, what if we imagined other philosophical projects as a new beginning to the conversation. A robot called Derrida arrives at construction sites and digs around the foundations so that they are not always covered up, maybe endangering the building. A robot called Nietzsche removes bits and pieces of the building to produce unwelcome gaps and erasures. A robot called Heidegger moves bricks carefully and mind numbingly slowly, taking a photo of each brick, and sending the image electronically to a live feed exhibited on a screen at the Museum of Modern Art in New York where afficionados are gathered watching and celebrating each brick (Figure 12.1). A robot called Zen arrives at the construction site and sits under a nearby tree for a year documenting the comings and goings of butterflies, insects, and foxes before allowing anyone to start. A robot called John Locke travels around the perimeter of the property and measures it again. A robot called Marx that needs a lot of oil only builds for non-profits while blaring out communist slogans. A robot called Richard Sennet arbitrarily

Figure 12.1 A robot named Heidegger. Author's composite.

hammers nails into planks while whistling a folk tune called "If I only were a Craftsman." A robot called Yunkaporta (an Australian Aboriginal philosopher) provides insights into indigenous practices while walking around the site on padded feet. A robot called Jung lays out a mandala using colored sand. A robot called Hegel builds during the day and unbuilds during the night, ultimately making a mess of things. A robot called Buddha only lays bricks when it is fed rice by local community members. A robot called Wittgenstein builds only if the future owner solves its word games. A robot named Kierkegaard only works if it is more than mile away from a church. It has a huge amount of moving gears and lays bricks with a shaky arm that perhaps even occasionally gets stuck. A robot called Sartre puts beams into the ground in arbitrary spots and then collapses in a heap.

All of this is, of course, quite silly, but it is no sillier than having to repeat Talos for the millionth time.

Notes

1 Shirley Darcus Sullivan, Sophocles' Use of Psychological Terminology Old and New (Ottawa, Carleton University Press 1999), 3.
2 *Iliad* 1.188-93.
3 Ruth Padel, *In and Out of the Mind Greek Images of the Tragic Self* (Princeton: Princeton University Press, 1992), 22.
4 Jason Horowitz. "Who Will Take Care of Italy's Older People? Robots, Maybe." The New York Times (New York, NY), March 25, 2023.
5 Jenna Owsianik. "State of the Sexbot Market: The World's Best Sex Robot and AI Sex Doll Companies." FutureofSex. February 11, 2022. https://futureofsex.net/robots/state-of-the-sexbot-market-the-worlds-best-sex-robot-and-ai-love-doll-companies/.

6 Mike Murphy. "I'm back at the grocery store and that dumb robot is trapping people who just want some creamer" Twitter, November 30, 2019 (2:17pm). https://twitter.com/mcwm/status/1200856424663961600.

7 Zack Estrin et al., writer. *Lost in Space*, Season 2, Episode 2, "The Precipice." Directed by Alex Graves, featuring Toby Stephens, Molly Parker, and Maxwell Jenkins, and aired on December 24, 2019.

8 "Members." The Unconventional Computing Lab, 2019. https://unconventionalcomputing.org/members/.

9 "The Greyhound bus lot on 934 Branna Street, now SOMArts Cultural Centre." SiteWorks: San Francisco performance 1969–1985, University of Exeter, 2013. https://siteworks.exeter.ac.uk/items/show/69.

10 "Machines." Roxy Paine. http://roxypaine.com/machines.

11 Tomorrow's Build. "This Robot-Built House Could Change Everything." YouTube, July 6, 2021, Video, https://www.youtube.com/watch?v=0wrQLDoJ6cw.

Bibliography

Jason Horowitz. "Who Will Take Care of Italy's Older People? Robots, Maybe." The New York Times (New York, NY), March 25, 2023.

Jenna Owsianik. "State of the Sexbot Market: The World's Best Sex Robot and AI Sex Doll Companies." FutureofSex. February 11, 2022. https://futureofsex.net/robots/state-of-the-sexbot-market-the-worlds-best-sex-robot-and-ai-love-doll-companies/.

"Machines." Roxy Paine. http://roxypaine.com/machines.

"Members." The Unconventional Computing Lab, 2019. https://unconventionalcomputing.org/members/.

Mike Murphy. "I'm back at the grocery store and that dumb robot is trapping people who just want some creamer." Twitter, November 30, 2019 (2:17pm). https://twitter.com/mcwm/status/1200856424663961600.

Ruth Padel. *In and Out of the Mind Greek Images of the Tragic Self.* Princeton: Princeton Unviersity Press, 1992.

Shirley Darcus Sullivan. *Sophocles' Use of Psychological Terminology Old and New.* Ottawa: Carleton University Press, 1999.

"The Greyhound bus lot on 934 Branna Street, now SOMArts Cultural Centre." SiteWorks: San Francisco performance 1969-85, University of Exeter, 2013. https://siteworks.exeter.ac.uk/items/show/69.

Tomorrow's Build. "This Robot-Built House Could Change Everything." YouTube, July 6, 2021, Video, https://www.youtube.com/watch?v=0wrQLDoJ6cw.

Zack Estrin et al., writer. *Lost in Space*, Season 2, Episode 2, "The Precipice." Directed by Alex Graves, featuring Toby Stephens, Molly Parker, and Maxwell Jenkins, aired December 24, 2019.

The Social

In 2009, architect and scholar William Mitchell introduced a special issue of A+U magazine on BIM by stating, "To place the emergence of BIM systems in perspective, it is worth recalling that architects' tools aren't neutral. They have particular affordances and biases, and they implicitly embody particular assumptions and values." Michell, who served as dean of the MIT School of Architecture and Planning from 1992 to 2003, was providing a kind of warning of techno-positivism: that despite the potential of the then-nascent technology, architects should be aware of its decidedly non-existent neutrality. While Mitchell used traditional tools—T-squares, triangles—as examples, the era of automation, big data, and artificial intelligence, has revealed many more technologically and socially complex "affordances and biases" related to architectural design tools. Today, software and hardware provide contexts for experimentation with different limits and potentials that even their creators are often unaware of. These range from explicit default settings to implicit assumptions about users.

We sometimes forget that engaging with technology is a social act. In architecture, we engage both people and machines in digital fabrication shops and computer labs. We troubleshoot our workflow problems on online forums. We build large scale prototypes in teams. The discourse, however, tends to focus on the end product or even the abstracted process. This can be seen, for instance, in the typical diagrams and high-fidelity animations of parts coming together seamlessly in empty voids that accompany so many digital fabrication research publications. Or, in the polished photos of uninhabited pavilions. Given that technological discussion in architecture tend to focus on either the

DOI: 10.4324/9781003296522-15

tool, the workflow, or the product, what is there to say about people in digital fabrication? What social effects has architectural digital fabrication had on architectural practice, education, and labor? What can we say about the spaces in which technology is deployed and who is given access to it?

This section is meant to shed light on the social dimension of digital fabrication in its various complex and messy forms. It takes on a notion of expanded research that extends outside of architecture into the worlds of sociology, labor, anthropology, ethnography, and what scholars in the field of Science and Technology Studies call socio-technics, the complex feedback loop between people and technology. Cross-disciplinary dialogues are necessary because architecture has long been an insular field. To understand the various social effects and relationships between people and tools we must look outside of architecture, perhaps even to science fiction and other literature. At the same time, architectural digital fabrication is different from engineering and manufacturing; it has its own system of values and perspectives. The discussions in this section borrow from these external disciplines as a way of revealing the affordances and biases of contemporary architectural digital fabrication. Moreover, they suggest ways to move forward with this in mind.

Construction Cheat Codes

James Coleman and Nathan Barnes

Disclaimer

The following requires a short disclaimer. This piece is being written from the perspective of a fabricator (subcontractor)—the owner of project means, methods, risks, and costs. In delivering architectural projects, we convert design intent into parts, budgets, and schedules. At Zahner, we also happen to be a sort of double agent, a mix of both architects and engineers. Like any good operative, we necessarily blur the boundaries between the two roles. For clarity, this case study will shift between our two worlds and perspectives offering *cheat codes* for navigating project delivery. Traditional project delivery maintains firm boundaries between participants (architect, engineer, fabricator, etc.) as a means of controlling risk and orchestrating liability. But this framework stifles communication between parties. Cheat codes offer creative ways for teams to work together better. Informed by digital fabrication, each proposed cheat code augments design coordination within existing delivery frameworks to create better solutions. Many of the suggestions for bettering project delivery are not sanctioned or in fact may be illegal per American Institute of Architects (AIA) contracts, situating them in a place between speculation and real-world applications.

Inherent ambiguity exists between design and delivery functions; pinpointing where one ends and the other begins is impossible. If one tried to pinpoint, they'd likely land in the realm of digital fabrication where design and making meet. To discuss what it

DOI: 10.4324/9781003296522-16

means to "home the machine" here, we'll illustrate best of both worlds scenarios where design and delivery efforts merge using a case study as an example of what each looks like. The focus will be on moments where manufacturing insights make designs richer and instances where designers can transform project delivery by participating in later stages of projects than they typically do.

Reorientation

> While the means and methods of construction are strictly speaking, the responsibility of the builder, a holistic approach to conceptual design cannot ignore implementation. Questions of sequencing, constructability, and tolerance are integral to a comprehensive design process.
>
> –Eric Höweler, Verify in Field[1]

Although digital fabrication has been available for over 20 years, inefficient, imprecise, and wasteful methods of building continue to dominate the larger Architecture, Engineering, and Construction (AEC) industry. This mode of working has led to increasing costs and degrading quality. We have more tools and technology than ever, yet our industry remains stagnant.[2] As this lack of productivity trend continues, ambitious design will continue to be viewed as too risky to pursue, pushing projects toward conservative or standard solutions.

If we hope to improve, we require a new approach. This new approach needs to begin with a redefinition and reorientation of what digital fabrication means to architecture. Instead of the current definition of digital fabrication as a singular act like cutting, printing, milling, and stacking, let us reconsider it as a series of interconnected physical and digital activities. There exists no singular fabrication process that transforms raw materials into complete buildings and no single group that holds the "distributed intelligence of the [AEC] community".[3] Buildings are the collective efforts of many different people and specializations, so we must consider the varying stakeholders throughout the process.

Craftspeople, factory workers, and field personnel are crucial stakeholders in digital fabrication. They are the holders of invaluable tacit knowledge and are also the downstream recipients of workflows, interfaces, parts, data, and decisions. New workflows and fabrication techniques must consider and optimize how it connects to and impacts these groups. In short, as an industry, we need to bring manufacturing expertise upstream and design delivery expertise downstream. To succeed in this, we'll need our cheat codes to maneuver around the silos and contractual barriers that halt communication. Success in this reorientation would look like a holistic digital delivery method that optimizes our collective efforts to deliver better design, reduce the penalty for variation (a.k.a., customization), and generate higher performance buildings more accurately at less cost. While a comprehensive process overhaul is likely warranted, in the meantime we present cheat codes for getting us from here to there.

Background: Off path

> ...digital fabrication machines are common machines used in many manufacturing facilities around the world; we just made them do things a little differently.
>
> –Bill Zahner, Digitally Defined Manufacturing

A. Zahner Company engineers have operated at the frontier of digital fabrication since the early 90s, contributing to architectural projects that push the limits of constructability in geometry, scale, and/or complexity. We work on roughly 50 projects simultaneously, each at varying stages of construction—from early conceptual design to final realized construction. This comprehensive practice enables Zahner to develop and deploy process improvements across the AEC industry.

In 2008, Bill Zahner wrote "Digitally Defined Manufacturing", nearly 15 years ago but also 10 years after delivering the ambitious Gehry Partners' Experience Music Project (MoPOP). Numerous advancements happened in the years following the completion of Experience Music Project (EMP) outlined in Bill's review of 10 projects completed between 1998 and 2008 including paperless delivery, collaborative coordination models, 3D model driven field support, and other ahead-of-their-time techniques.

These pioneering efforts cut the path for delivering extraordinarily complex projects and, perhaps more importantly, demonstrated the power of digitally fabricated architectures. And yet for many architects and AEC professionals, Bill Mitchell's 2001 writings (Mitchell) on "Digital Delivery" and "CAD/CAM translation of design intent" read as if they could be published in 2023.[4]

> Do you do the "BIM"?? - industry experts

The realities of our risk-averse industry can make executing novel workflows difficult. Architects can provide real value to later stage stakeholders like estimating, factory, and field personnel but many barriers exist between groups. By playing a larger role in the shaping of project means, methods, and evaluation, architects can use their technical dexterity to tailor communication between groups and guide project conversations while preserving design intent.

Case Study

Using Zahner's work on a recently completed tech campus (Figure 13.1)—designed collaboratively by Heatherwick Studio and Bjarke Ingels Group with Adamson Associates as Architect of Record—we will demonstrate moments of seamless delivery on a large-scale audacious project. The project's novel design and scale made it a good candidate for a reevaluation of delivery methods, which were at odds with conventional construction processes. On the project, we faced challenges with estimating, fabricability, design

Figure 13.1 Case study project, tech campus (project identity withheld) with stainless steel clerestory eyebrows and perimeter bullnose—designed, engineered, fabricated, and installed by Zahner.

communication, and field installation. Crucial to our success was the use of process cheat codes which we think are a useful starting point for reconsidering project delivery overall.

Zahner contributed to the campus by engineering, fabricating, and installing two scopes of work: the clerestory "eyebrows" (1,076 units and 3,620 coping panels) and a perimeter bullnose (232 - 25′ long assemblies). The clerestory was made up of a network of interconnected ~100′ long doubly curved ribbons (curved in plan, elevation, and section). The similarly shaped undulating bullnose wraps the perimeter of each building, enclosing the roof edge, and supporting a perimeter gutter. We will focus on the clerestory scope shown in Figure 13.1.

The various project cheat codes outlined below are specific strategies that emphasize digital fabrication not solely as a form of making but also as a form of thinking, planning, and designing. They include software to support better estimating, prototyped fabrication processes that establish parameter space for designs, methods for optimized information exchange between design and production groups, and the use of 3D models to track and coordinate work at the jobsite. Collectively, these cheat codes paint a picture of a more holistic collaboration among designers, engineers, craftspeople, and machines.

Cheat Code #1: Make Buildings Easier to Estimate

As a project matures from concept to construction, designs need to be costed and reviewed for feasibility. Cost is perhaps the most important parameter in a project as it determines what is included or excluded. However, architects often don't have enough information to know what design decisions drive costs. Therefore, they must initiate a dialogue with industry experts to evaluate and optimize the design. Iteration underlies such optimization, so the more times the design-to-cost feedback loop between architect and subcontractor is completed the better the result. When cost is built into a parametric model exploration of form and performance happen alongside cost estimation. We fabricate most of our parts using Computer numerical control (CNC) cutting machines that have well-understood parameters for cutting speed, cycle times, etc.—so collating geometry information with cost information is straightforward for Zahner.

To accomplish this, the stages of design and estimating must be well coordinated, which in most cases is simply better organization of design information. When alignment between design and estimating teams happens, design iterations can be more numerous, maximizing design while minimizing cost.

It is all too common to have design documentation that is unclear or incomplete. When this happens, estimators make conservative assumptions that result in elevated costs. When costs come in high, the construction manager begins looking for cost savings through value engineering, which often leads to needless design degradation.

Bid packages today are typically clumsily collated data dumps due to the volatility of concept design and varying software, file types, methods, and competency of the design team. The results of this haphazard documentation effort are mashup models of meshes, duplicates, clashing elements, and extraneous geometry that require herculean efforts by estimators to comprehend and quantify. If a design is difficult to convert into data, estimating teams will spend most of their efforts interpreting documentation rather than optimizing costs. Since subcontractors are not typically paid for estimating and pre-engineering (nor are they guaranteed they will be awarded the project) such efforts are deliberately limited.

Here, we can learn from digital fabrication methods of organizing model components. Just as you might generate a drawing template or assemble specific "Etch", "Cut", and "Score" layers in your laser cut file to save time during cutting, similar methods can impact how your design information is processed during the costing phase.

During pre-construction on the case study project, we employed an example fluid model review technique for the clerestories as it underwent iterative estimating. We worked closely with the Architect of Record to organize and share 3D models of the project at a regular cadence. Zahner was just one of many subcontractors who worked on the canopy. As multi-trade feedback (steel, glazing, drainage, etc.) was incorporated into the process, the design necessarily changed as the design team optimized for cumulative constructability, cost, and performance. To track design changes and verify feasibility of each iteration, we automated the creation and evaluation of 1000's of cross sections using the software tools Grasshopper and MS Excel (Figure 13.2a). With each new iteration, we automatically evaluated and verified that constraints (e.g. available raw material sheet sizes and structural limitations) were not broken. The result of these design checks was then exported from Grasshopper and visualized in a tabular data format. When design changes broke our constraints/assumptions and risked cost increases, we elevated the issue to the design team for resolution. This collaboration was critical for project success. The tools allowed us to quickly evaluate and communicate cost and design implications which enabled the design team to make informed decisions. The restructuring of 3D geometry into tabular data is trivial for design tools like Grasshopper but can leave subcontractors flummoxed—architectural drawings are critical for communicating design intent but can be useless for quantifying a complex project.

Cheat Code #2 Mockups: Verify You Can Make (And Assemble) What Your 3D Model

Not every 3D model is buildable. Complex designs require physical testing to validate that they are realizable. When digital representations match reality, the model can drive production and quality control processes in a clear path from design to construction. If done well, everything fits, design intent is met, and parts are produced in a linear process.

Any building component that is 3D modeled, exported as a 2D file boundary, and cut from a sheet on a modern CNC machine should be perfect (within reasonable construction tolerances). Unfortunately, not all parts of a building maintain this level of precision, many are further transformed (bent, rolled, welded, etc.), which can lead them to diverge from their modeled basis due to varying material properties and/or process tolerances.

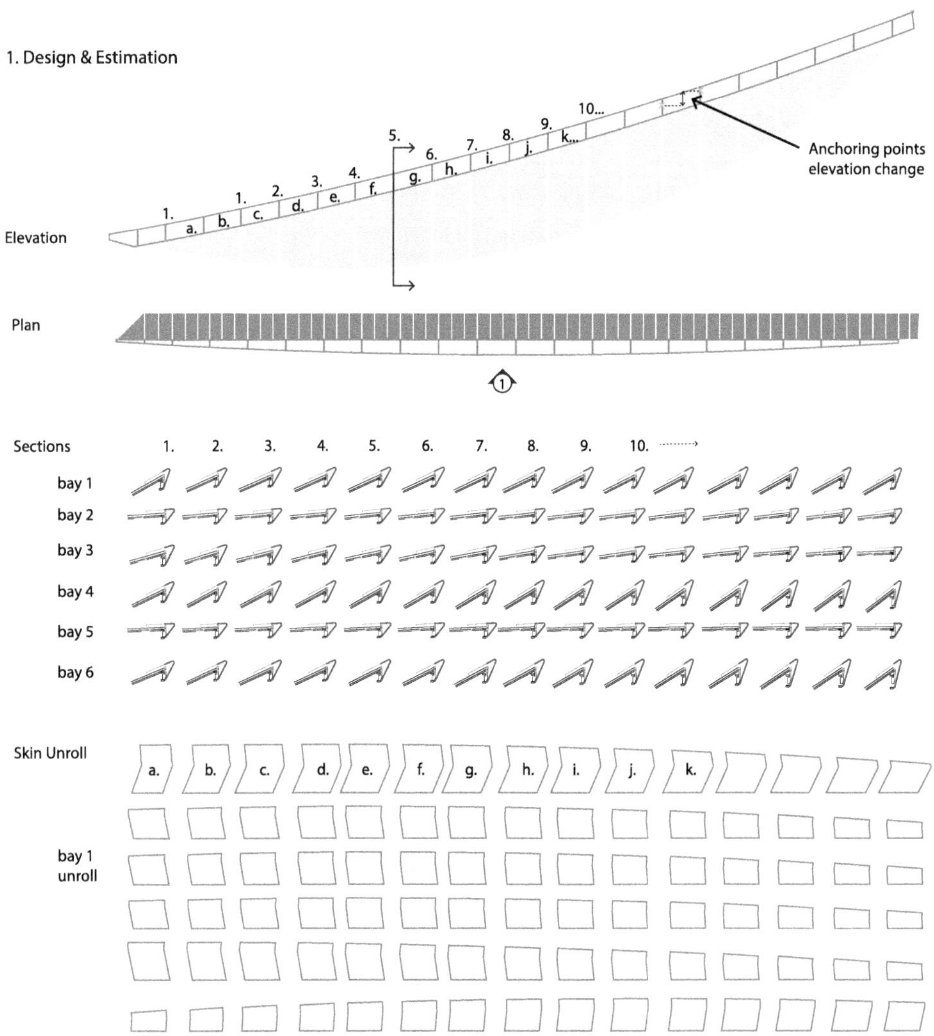

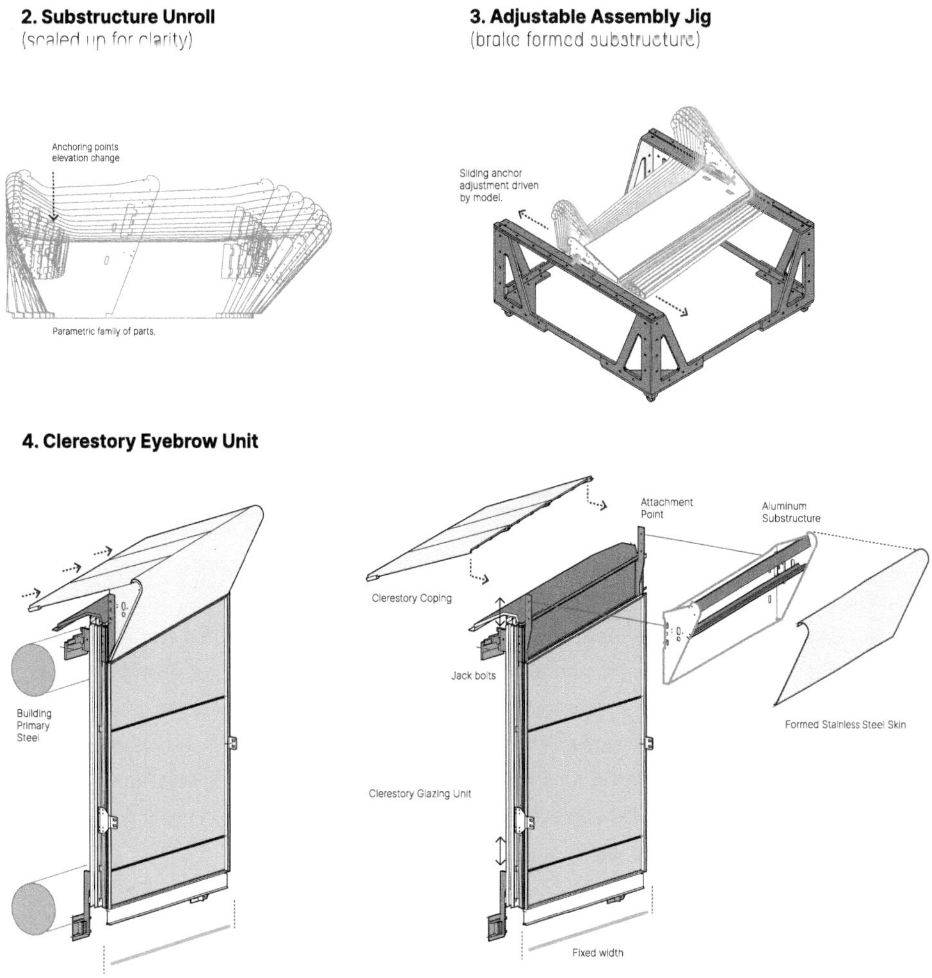

Figure 13.2 a and b Diagrams describing clerestory design, workflow, and fabrication strategy.

Design intent models are a starting point—from there the design needs to be redefined to meet criteria of budget, performance, fabricability, installability, schedule, etc. To answer these questions, expertise from many different functional groups and specializations is needed. But this crucial knowledge is often missing during traditional design phases. To ensure a design is buildable, 3D models must be calibrated via targeted empirical testing with craftspeople to establish parity between digital and physical assemblies (Gelpi). At Zahner, engineers established a bi-directional exchange in which feedback from the shop on how materials behave (e.g., stretch out), informs design parameters, and guides how parts should be modeled and drawn.

As originally designed, the clerestory ribbons were monolithic doubly curved surfaces composed of a morphing cross section swept along two arcs (Figure 13.2a). The radiused surface required rationalization to become fabricable. Our solution for forming the skins was to use a custom engineered brake press die. After successful forming

Figure 13.3 Brake forming process for stainless steel skins using custom tooling. This approach was chosen because of its flexibility and ability to form multiple radius cylinders and conic forms.

tests, process parameters were communicated to the design team and inserted back into our digital model. Info added to the model included rules on where radiuses needed to start and stop, values for material deformation, and how the surfaces needed to be 3D modeled per the tooling and forming process tests (Figure 13.3).

When mockups were assembled, Zahner craftspeople made notes on helpful additions that could be included during 3D modeling that could simplify the task of fabrication. These included labels, notches for alignment, indexing features, scribe marks, etc. that could be easily added by the CNC machine while processing a part. This is what is meant by thinking through the lens of digital fabrication. A feedback loop among designers, engineers, and craftspeople leads to digital fabrication choreographing activities well beyond the CNC operation.

Cheat Code #3 Parametric Production: Consider What Kind of Information Is Most Easily Interpreted and Verified. Check Lists and Jigs Are Better Than Drawing Sets

During the design for manufacturing phase, projects are often needlessly standardized (*watered-down*) because of a perceived risk associated with part variation. When delivered using conventional methods, part variation can require elaborate drawing sets, which can be burdensome for designers and fabricators. This does not have to be the case as there exists a method for achieving variation without raising costs. If rule-based processes are carefully considered during design, there is a best of both worlds scenario. The key is to quantify variation of a design and develop a method for communicating and controlling variables throughout the assembly process.

"Jigs" or "fixtures" are custom elements made to aid in the manufacturing and/or assembly of a product—used to increase repeatability and simplify production. They take many different forms from clamps to frames or custom hand tools, used to lower the complexity of a task and increase quality through repeatability. Unfortunately, the production efficiency afforded by typical jigs in a standardized process comes at the cost of flexibility and limits potential design solutions.

In contemporary construction situations, this means that design elements such as filleted corners, daring twisted façades, perfectly patterned elevations, and doubly curved geometries, when viewed through the lens of standardization, will be marked as expensive, impossible, or impractical. However, when jigs are paired with a parametric model, they can enable part variation and reduce efforts required of the craftsperson. There is a class of jigs/fixtures that include adjustable and or interchangeable components (Henriksen) that allow for a single jig to output multiple part types (or a parametric family). In this scenario, numeric parameters from a 3D model can drive the arrangement of adjustable jigs, allowing parts not only to vary but also be verified with minimal effort and no drawings.

As mentioned, geometric variation can be costly if it requires extra work for engineers to generate one-off directives and increased effort in the factory for craftspeople to interpret nonrepeating instructions. When no standards exist, quality control efforts must be also multiplied to verify that the output matches the varied part dimensions. Conversely, when a product is standardized, a process can be streamlined to reduce production efforts.

> The obvious place for jigs and fixtures is in mass production, where large quantity output offers ample opportunity for recovery of the necessary investment.
>
> Henriksen

In our case study, the clerestory units varied widely. They were composed of a radiused stainless steel skin mounted to a formed aluminum substructure (coined the 'batwing') to create a segment of the clerestory ribbon. The units were mounted directly to the curtainwall via bolt-on brackets that extend upward outside the primary weather barrier (Figure 13.2b). While each of the units was unique in length, orientation, and cantilever, they were constructed from the same rule set, a parametric "family" (Figure 13.2b). The elevation of the clerestory anchor points was determined by the slope of the ribbon over the width of each glazing module—as the slope changed the anchor points slid up and down in elevation. To ensure fitment of the eyebrow units with the corresponding glazing anchors, we fabricated a parametric jig (Figure 13.4) with sliding anchor points that replicated all real-world conditions. The clerestory jig acted as both a quality control check and assembly station—a model-driven calibration point in our production process.

The adjustable jigs developed for the case study were crafted around the rules of our parametric model and driven by a spreadsheet of values. This pairing made geometric validation straightforward without a lot of drawings, achieving variation without burdensome cost. We viewed the clerestory design through the lens of craftsperson effort and jig design, streamlining manufacturing coordination while preserving design intent (Figure 13.5).

Figure 13.4 Adjustable clerestory assembly jig with brake formed "batwing" substructure mid-assembly. The jig allows for varying part dimensions driven by parameters pulled from the parametric model.

Figure 13.5 Clerestory mockup of a complex roof intersection at network node. Lessons learned during mockup fabrication were reinserted into the larger design and modeling process.

Cheat Code #4 Field Support—Help Collect, Model, Visualize, and Synthesize Data On-Site

3D modeling is often considered a task that takes place exclusively during the planning phases of a project. This is a missed opportunity and misconception. 3D models are critical during planning phases but we argue that they are even more critical during execution phases when used as a means of understanding the state of a project and directing activities. The field is the riskiest and most expensive phase of any project, models should be used to analyze building conditions and adapt strategy on the fly.

A construction site is a volatile and ever-changing place. As a result, determining the state of a project at any given moment is a non-trivial task. Many parts, especially big heavy steel elements or phase-changing materials like concrete, often drift from the theoretical target shape and size. Designs must take dimensional deviations into account, and fabricators are allowed a certain amount of tolerance based on a reciprocal relationship between cost and precision. On most projects, our tolerance takes the form of slots, shims, and field drilled holes. Based on conditions on-site, these features allow us to install our parts in the correct position even when predecessor work might be out of position.

Positional flexibility provided by tolerance is crucial for any installation. It is even more critical for prefabricated parts (like the ones Zahner produces) because they are not mutable once on-site. The question becomes how to best use the tolerance available. Systems with lots of tolerance provide a lot of options but it becomes difficult for field installers to determine which solution is best from the solution set available.

As a building is constructed, it exists in numerous states, growing/shrinking with temperature swings, deflecting, and translating as loading changes. This makes it difficult to know where anything is at any one point in time. On-site personnel own the difficult task of determining the project "state" and subsequent determination of what corrections are necessary. This challenge of interpreting project conditions is an opportunity for design teams to support on-site activities. 3D-modeling tools common during the design phase (such as Rhino/Grasshopper) can be used to analyze the state of an installation, determine where adjustments should be made, and feed information back to teams on-site, saving time and increasing quality.

The clerestory glazing assemblies in the case study were hung from steel angles welded on the primary structure. The assemblies had ±1.75" of vertical adjustment via jackbolts and several inches of left/right adjustment to accommodate steel fabrication tolerances. Three points on every bracket were surveyed prior to install. Instead of reviewing 100s of PDF reports describing the surveyed points (state of each bay), we requested the Surveyor provide us with the raw point data from the total station. With the point data in a .txt format, we could read the points into our Rhino model using Grasshopper and Python scripts to plot the measured positions in 3D space. By coordinating directly with raw positional data instead of paper documentation, we were able to interpret the state of the anchors and respond more quickly. Additionally, once plotted in our Rhino model we could analyze the data more robustly than can be done using traditional periodic PDF survey reports. In the case of the clerestories, we incorporated the canopy deflection criteria in our evaluation and optimization (translations in X, Y, Z as the building was loaded with more weight) to predict where our final position would be and respond appropriately (Figure 13.6). This analysis resulted in our setting of units strategically high so that once the roof was complete the units would settle into the correct location. The building was in a constant state of change, so deflection data needed to be

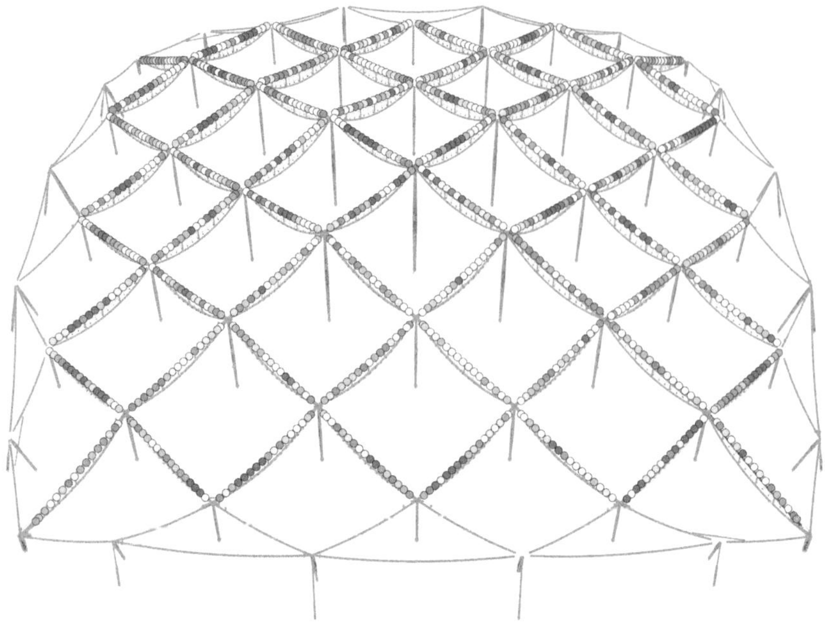

Figure 13.6 Map showing plotted survey points generated in Rhino/Grasshopper, illustrating positional variation due to error accumulation and deflection criteria.

prorated for the portion complete. Using the automated tools we built, this adjustment was as simple as checking a few boxes for what layers of the roof had been installed.

Once all the analysis was complete and positional data calculated, a spreadsheet of elevation values via jackbolt heights was exported and sent to the field crews via a shared cloud folder. This offered the installers the ability to set the positions of all the clerestory units while safe on the ground (Figure 13.7) with assurance that parts would align once the building was fully loaded (Figure 13.8). When compared against the initial scheduled time, we were many times faster than planned. When considering the cost of occupying a tower crane or deploying crews on-site (not to mention the reduced project risk by lessening the time installers were in dangerous positions on the roof), the benefits were tremendous. The installation of the complex clerestory assemblies is just one example of how digital fabrication-based thinking and strategizing combined with model-driven installation can support field personnel to deliver higher quality at lower costs in safer ways.

Conclusion

Outlined in the above case study are ways in which architects, craftspeople, and various stakeholders can not only build more efficiently but also more precisely and safely. The cheat codes we developed for estimating costs, understanding and sharing manufacturing constraints, and doing complex field installations embody a new type of creative

Figure 13.7 Zahner install teams setting jackbolt heights of curtain wall assemblies on the ground prior to crane hoisting. Setting information was calculated and optimized in an automated Rhino/Grasshopper workflow.

Figure 13.8 Results of initial "rough setting" of clerestory assemblies. Success of survey analysis was validated by plumb alignment of assemblies despite varying anchor positions.

Figure 13.9 Final roof assembly of clerestories and coping panels along canopy ridgeline. Clerestory node intersections are visible in distance.

thinking informed by digital fabrication. The techniques described in the case study worked very well for us as we executed the design intent on a large complex project, but the lessons are extensible. This kind of effort is needed not just in projects similar to the case study but across all AEC work.

To get there, we require a wider perspective on digital fabrication in architecture. We must consider the integration of new technology and architectural research into construction and the impact on downstream efforts. Let us consider the entirety of the process as a digital fabrication project and look for ways to collaborate across disciplines (sanctioned or not). Research in this emergent field will then have a better chance of industry transformation because it will be rooted in and informed by the people whose hands craft our buildings. We hope that by optimizing our collective efforts we don't simply build faster but also spend the saved time endeavoring to design and build better (Figure 13.9).

The closed-loop control systems underlying most digital fabrication machines maintain positional accuracy by continuously comparing the current position (the state) of the machine with the target position described in the machine code. This is done at a high frequency so highly accurate parts are possible via subtle corrections made hundreds of times per second. Much of what we've described above are methods of communication, and feedback loops between varying groups that describe a state and keep projects on track. Unfortunately, sharing between groups is easier said than done in a high risk and litigious industry. Ideally, coordination between groups could be achieved robustly through APIs and seamless data exchange—a good goal but one that would require a restructuring of the entire industry—for now we offer these cheat codes.

Proving out a novel technique or cheat code can be tricky, but the burden of proof lies with the proposer. Earning the trust of a skilled craftsperson or project delivery expert takes time, patience, and empathy—if they are resistant to change—it may be because your proposed solution is making their already very difficult job even harder!

Notes

1 Howeler, Eric. *Verify in Field: Howeler + Yoon Architecture*. Zürich: Park Book, 2021.
2 Changali, S.; Mohammad, A.; van Nieuwl, M. *The Construction Productivity Imperative*. New York: McKinsey & Company, 2015.
3 Pittman, Jon. "Building information modeling: current challenges and future directions." *Branko Kolarevic Architecture in the Digital Age: Design and Manufacturing*. New York: Taylor and Francis (2003): 254–258.
4 Mitchell, William J. "Roll over Euclid: how Frank Gehry designs and builds." *Frank Gehry, Architect* 354 (2001).

Bibliography

Changali, S.; Mohammad, A.; van Nieuwl, M. *The Construction Productivity Imperative*. New York: McKinsey & Company, 2015.
Gelpi, Nick. *The Architecture of Full-Scale Mock-Ups: From Representation to Reality*. London: Routledge, 2020.
Henriksen, Erik Karl. *Jig and Fixture Design Manual*. New York: Industrial Press, 1973.
Howeler, Eric. *Verify in Field: Howeler + Yoon Architecture*. Zürich: Park Books, 2021.
Mitchell, William J. "Roll over Euclid: How Frank Gehry designs and builds." *Frank Gehry, Architect* 354 (2001).
Pittman, Jon. "Building information modeling: current challenges and future directions." *Branko Kolarevic Architecture in the Digital Age: Design and Manufacturing*. New York: Taylor and Francis (2003): 254–258.
Zahner, William. "Digitally Defined Manufacturing." Kolarevic, Branko, and Kevin Klinger. *Manufacturing Material Effects: Rethinking Design and Making in Architecture*. London: Routledge, 2013.

Eco-Digital Construction in Africa

Decolonizing Performative Architecture from the African Fabbers Project to the Douala Cultural Hub

Paolo Cascone and Maddalena Laddaga

Introduction

This chapter is a discussion of the experimental work of CODESIGNLAB based on more than ten years of realized projects and applied research from the African Fabbers Project (AFP) to the construction of the Douala Cultural Hub. The work investigates the relations between African indigenous technologies and advanced design approaches toward synthetic vernacular architectures. Access to education and the development of creativity represent to us some of the major drivers for improving the life quality of a community from both social and environmental points of view. With this premise, as designers, educators, and activists, we have conceived of our architectural projects as creative open platforms: a set of urban infrastructures able to provide spaces of production for both educational activities and community-oriented projects.

The Making of a School of Self-Construction

At CODESIGNLAB we have started to explore new ecological and eco-digital construction approaches through the development of the AFP, a set of case studies realized in different cultural contexts around Morocco, Ghana, Mali, Burkina Faso, and Senegal. The project has been shaped to bridge informal and advanced fabrication approaches with the aim to reshape public spaces into spaces of community and production through

DOI: 10.4324/9781003296522-17

collaborative practices. Our approach is based on the exploration of new paradigms on circular economies and "local industry" potentials for African cities' post-colonial growth.

By 2050, Africa's 1.1 billion citizens will likely double in number, and more than 80% of that increase will occur in cities, especially in high poverty communities (UN Habitat). The problems of African societies are often depicted by Western media as just an object for humanitarian aid. However, as explained by major international analysts, three powerful positive trends are likely to characterize Africa's growth in the coming years. First, as mentioned by the United Nations Development Programme (UNDP), the continent has a young population with a growing labor force—a highly valuable asset in an ageing world. Second, Africa is still urbanizing, and much of the economic benefit lies ahead. Productivity in cities is three times as high as in rural areas and, over the next decade, an additional 187 million Africans is projected to live in cities, according to the United Nations. Third, African economies are well positioned to benefit from rapidly accelerating technological changes that can unlock growth and leapfrog the limitations and costs of physical infrastructures in important areas of economic life. Based on this premise, the case studies to follow will be compared and analyzed with the aim of developing a possible design-build processes that can bridge vernacular, spontaneous, and advanced cultures. This, in turn, can generate new opportunities of self-organization and self-production for the African urban context.

A key project that embodies this potential for the AFP was the African Fabbers School in Douala, Cameroon. Thanks to the support of the Italian Ministry of Foreign Affairs, we were able to establish the first community-oriented school of urban ecologies and self-construction in Africa. This educational and research platform is based on a learning-by-doing methodology that investigates the new social role designers can take as facilitators of participatory processes for regenerating residual public spaces within a circular economy approach. This methodology was the result of a series of on-site urban experiments developed in the last 15 years with the aim to deploy different urban tactics according to site-specific conditions and social scenarios.

Learning from Pioneers

Our approach has been inspired by the collaboration between Paolo Cascone and Fabrizio Carola, an Aga Khan award-winning architect known for his pioneering 30 years of work in the sub-Saharan region. Carola is an independent thinker, one of the few who have been able to develop theory via practical work, anticipating contemporary themes crucial for new generations and for more than just architects. He quickly understood that Africa was the right place to test an ecological approach to urban design and architecture where social, mental, and environmental ecology could flow into one another. He understood well that technology would play a fundamental role in making building processes sustainable. Above all, he understood—before many others in the field—how important it is to work in a state of osmosis with the landscape and the people living there with a post-colonial perspective. From this point of view, his interdisciplinary approach can be seen today as very advanced, despite the low-tech techniques used. In his work, the building

Figure 14.1 Fabrizio Carola in Mali.

site was also his studio, an applied research laboratory focused on an idea critical in Africa: self-sufficiency. Using this simple concept, he developed a method that could generate concrete responses to problems created by climate change, such as deforestation and impoverishment. Through a simple yet integrated process, he transformed the critical environmental and economic issues of an urban context into generative elements in his designs. Equally, his tendency to see the construction site as a place of production and dissemination gave his work a subversive character. His idea of evolving Hassan Fathy's compass to create building shells (such as domes and vaults) entirely using local or "0 km" materials is one example of his innovative solutions (Figure 14.1).

Furthermore, our agenda was also gradually informed by other interdisciplinary approaches on post-colonial issues, such as the work of Pier Paolo Pasolini in his documentary "Notes Towards an African Orestes" (1970) and the work of Rem Koolhaas: "Lagos, How it works" presented at the Venice Biennale 2010. While Pasolini's documentary has a critical and somewhat actual description of the controversial development of new post-independence African cities and society during the seventies, in the case of Koolhaas' research on Lagos (developed with his students of the GSD Harvard), the emphasis was put on the consequences of this African modernism and the capability of informal activities to shape the so-called generic city. Another interdisciplinary approach relevant to our research comes from the work of the ethnic-mathematician Ron Eglash at the intersection of computer science and cultural anthropology. In his book African Fractals, Eglash describes the sub-Saharan spontaneous settlements as emergent systems coming from implicit and so-called contextual algorithms. Such spontaneous systems analysis explains in a scientific way the cause-effects relations among complex geometric configurations, functional aspects, and social organizations. These interactions generate resilient solutions responding to very interesting criteria of material optimization, spatial diversity, and environmental responsiveness.

The Making of a School of Eco-Digital Self-Construction

With this cultural background, the agenda of the AFP started to develop into an applied research platform that would use design to make strategies for the African urban context with the aim of generating new urban *synthetic vernacular* prototypes. The first experiment was developed with Atelier Paolo Cascone at the École Spéciale d›Architecture (ESA) in Paris in collaboration with the local community of Ouagadougou in Burkina Faso. The design process aimed to develop a high-tech design/low-tech construction approach to performative design; it merged computational processes with vernacular construction techniques around the use of earth bricks for evolving Nubian vault technologies. The project is a direct result of the participatory process involving the community of Sourgoubila in building a primary school playground. This space is conceived to allow the children of the school to play and study in a shaded playground (Figure 14.2).

This on-site initiative then allowed us the opportunity to develop the smart shell project at the 6th World Urban Forum. This microstructure was conceived in order to provide off-grid shells for both public activities and temporary housing solutions for African cities. The shell is designed as a transportable, fast-deployment, and energetically self-sufficient structure. The design process investigated the relationship between high-tech design and low-tech construction. The project provides flexible solutions for different uses, from low-cost dwellings to off-grid emergency hospitals and so on.

Figure 14.2 Earth[play]ground, construction process, Burkina Faso.

Figure 14.3 Smart Shell project.

The structural system consists of foldable modular arches made of wood and covered by photovoltaic textile strips (Figure 14.3).

Another on-site experience of the African Fabbers program in Morocco was in the city of Sale in Morocco. This project was developed in the framework of the Quartier du Monde initiative in collaboration with the Swiss Agency for Development and Cooperation to involve the inhabitants of a poor neighborhood in public space regeneration. The lightweight structure we designed is conceived as a tensile fabric shelter for the public garden of Sale. The project was inspired by the traditional cable system for generating shading solutions in the ancient medina. The aim was to create a community space in cooperation with the local community involving them in a self-construction project for a common shaded gathering place (Figure 14.4).

After this case study, the AFP was invited to participate in the Marrakech Biennale in 2014. The project was a transformation of an abandoned hangar into a productive atelier for Moroccan design students and the local community of artisans. The space was equipped to be a temporary laboratory with a big delta 3D printer for natural materials, an Arduino coding location, and a computer area. Inside the atelier, we developed together with the local artisans an energetically self-sufficient 3D printer and explored the potential of 3D printing with natural materials to realize structural components prototypes for a hydroponic pavilion made with interlocked ceramic elements fabricated by extruding local clay. With this, both urban and educational experience, we started to be in dialog with the network of African fab labs, helping them become urban fabrication laboratories (Figure 14.5).

A significant project that further developed this analog-digital approach in manufacturing was realized for the Dakar Biennale in 2014. This initiative was developed to support the DefKo Ak Niep (Do it With Others) fab lab in Senegal. The project provides a small infrastructure and an open-air laboratory for local craftsmen and makers. The structure's construction has been designed and realized through a participatory process involving both the local community and European makers to bridge traditional and advanced knowledge. All of the above-mentioned experiences convinced the AFP team that an innovative education system was needed in order to respond to the lack of universities and education programs in the region and support a new generation of urban drivers (Figure 14.6).

Figure 14.4 Khaima urban tent self-construction, Morocco.

Figure 14.5 Marrakech Lab, Morocco.

Figure 14.6 Open-air lab, Senegal.

The Douala Cultural Hub Project

In 2017, after many years of itinerant sessions in sub-Saharan Africa, where we developed the methodology and the knowledge exchange program, the African Fabbers School moved to Cameroon for a few years. In 2021, a new Cultural Hub was inaugurated in the city of Douala, Cameroon. The project was designed by our architecture office Cascone + Laddaga/CODESIGNLAB for the COE NGO with the support of the Italian Agency for Development Cooperation. The construction of the hub, displayed at the 2021 Venice Biennale of Architecture as part of the "African Fabbers School - AFS" installation, was the result of a collaborative process that involved the students at the school, a laboratory of social design and self-construction, as well as the local community. The main drivers of the design process developed by our office aimed to respond to the extreme climatic conditions of Douala as well as the necessity of a circular economy approach that prioritized local materials and construction knowledge. For these reasons, the cultural center was parametrically designed with an environmental approach and built by bridging traditional techniques and digital fabrication technologies. The use of advanced tools of environmental analysis informed the ecological design process to minimize the energy consumption of the building, thereby improving its passive cooling ventilation and thermal performances. The project was conceived as an urban micro infrastructure that includes a school of fine art, exhibition spaces, laboratories, and an open structure built around a courtyard that is able to evolve over time according to the local community's needs.

Furthermore, since 2017 we have strategically started to map the context of the school in terms of local knowledge and existing networks of artisans and creative communities. Our goal is to create one of the first self-production laboratories for digital and traditional manufacturing for sustainable living aimed at transforming local materials such as wood, clay, and so on. We have also carried out workshops with groups of local artisans and students, recruited through open calls published on social media, and trained over time by generating community-oriented micro infrastructures. The school has also

Figure 14.7 African Fabbers School, Douala, Cameroon.

Figure 14.8 African Fabbers House, Douala, Cameroon.

implemented an off-grid housing research program geared toward affordable dwellings and upgrading informal settlements. The program looks at scalable off-grid housing solutions by bridging vernacular architecture with more advanced technological systems inspired by Jean Prouvé's tropical house project.

African economies are well-positioned to benefit from rapidly accelerating technological changes. They can unlock growth and leapfrog the limitations and costs of physical infrastructures in important areas of economic life. This strategic growth is crucial for developing local next-generation industries that will transform and repurpose local materials within an ecological approach that reconciles nature with tekne' (Figures 14.7–14.10).

Figure 14.9 African Fabbers House, Douala, Cameroon.

Figure 14.10 African Fabbers House, Douala, Cameroon.

Notes for the Future

Thanks to the opportunity to develop the African Fabbers installation for the Venice Biennale in 2021, we have started working on the AF Atlas book for Actar publishers as a way to reconstruct the methodological fil rouge behind each of these experiments. This diachronic approach intends to generate, after many years of post-colonial studies, an operative agenda of possible strategies, which is in accordance with different conditions starting from the anthropological and the climatic. Such an agenda responds to a global need for an ecological shift in contemporary design and manufacturing processes, which should bridge high- and low-tech cultures.

This work has generated a manual that is articulated around emergent principles inspired by traditional and informal African practices and architectures: self-similarity and diversified typologies, material optimization and circular economy, self-sufficiency, and responsive dwellings (Figure 14.11).

This climate/anthropology-driven approach has even created a taxonomy of inclusive tectonics: open systems that are able to adapt themselves and regenerate public spaces through collective practices. Beyond the many built projects and experimental protocols explored in the African Fabbers initiative, and thanks to the contributions of hundreds of students, artisans, and other participants, this experience has radically changed our research agenda as well as our cultural perspectives. As educators, designers, and researchers, we systematically learn from informal practices as modus operandi to address complex dynamics also in Western countries, anticipating the ongoing debate on decolonization in architecture. The idea that we are now developing with our partners is to shape a more structured knowledge platform between Europe and the Global South in order to adapt and test the same methodology for other climatic and social contexts (Figures 14.12–4.20).

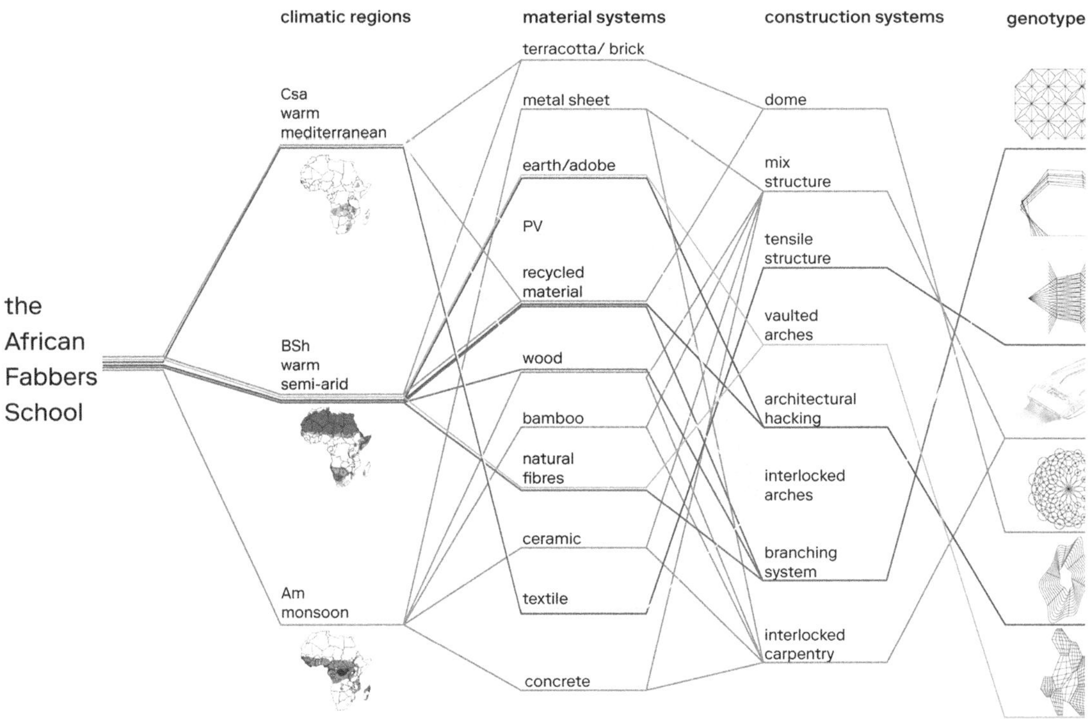

Figure 14.11 African Fabbers School projects taxonomy diagram.

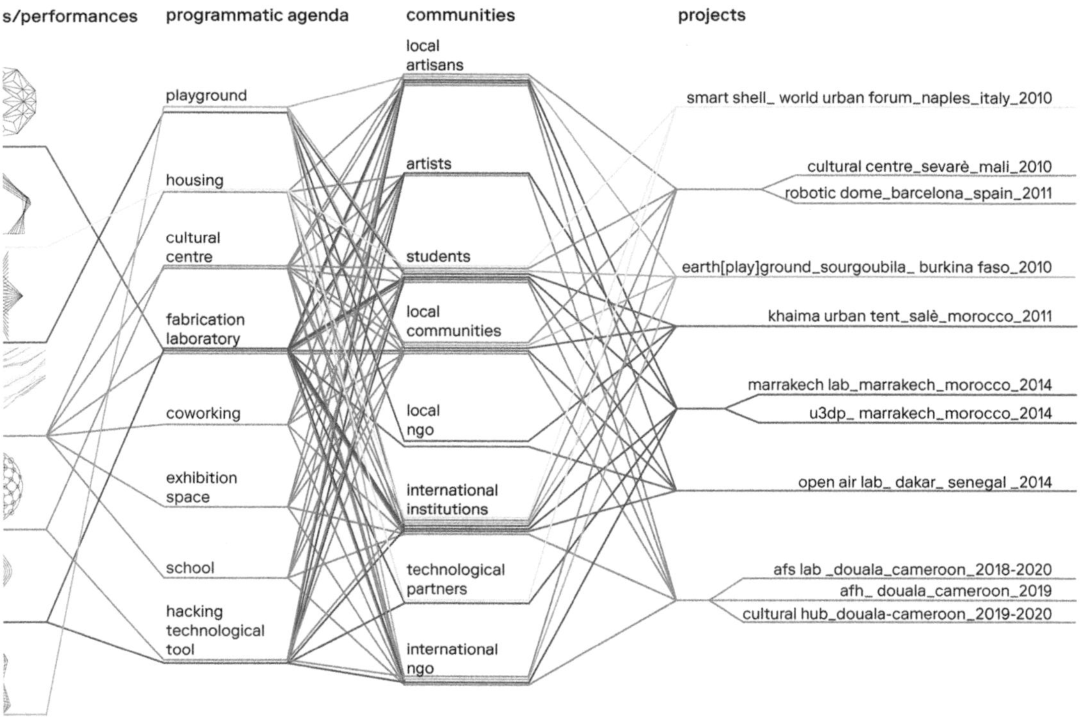
s/performances
programmatic agenda
communities
projects
playground
housing
cultural centre
fabrication laboratory
coworking
exhibition space
school
hacking technological tool
local artisans
artists
students
local communities
local ngo
international institutions
technological partners
international ngo
smart shell_ world urban forum_naples_italy_2010
cultural centre_sevarè_mali_2010
robotic dome_barcelona_spain_2011
earth[play]ground_sourgoubila_ burkina faso_2010
khaima urban tent_salè_morocco_2011
marrakech lab_marrakech_morocco_2014
u3dp_ marrakech_morocco_2014
open air lab_ dakar_ senegal _2014
afs lab _douala_cameroon_2018-2020
afh_ douala_cameroon_2019
cultural hub_douala-cameroon_2019-2020

Figure 14.12 Douala Cultural Hub, design process, Cameroon.

Figure 14.13 Douala Cultural Hub, wood façade self-construction, Cameroon.

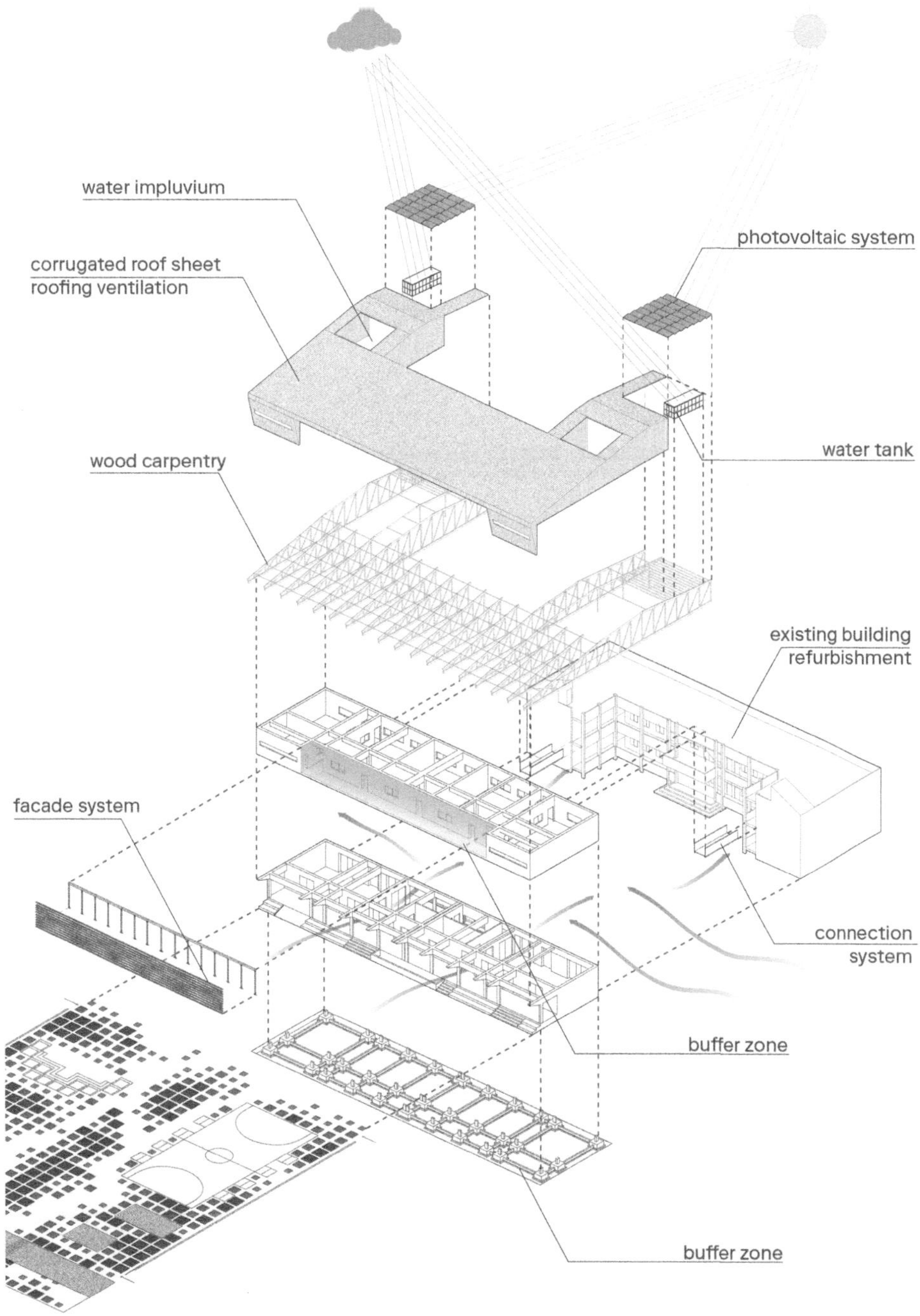

Figure 14.14 Douala Cultural Hub, diagram, Cameroon.

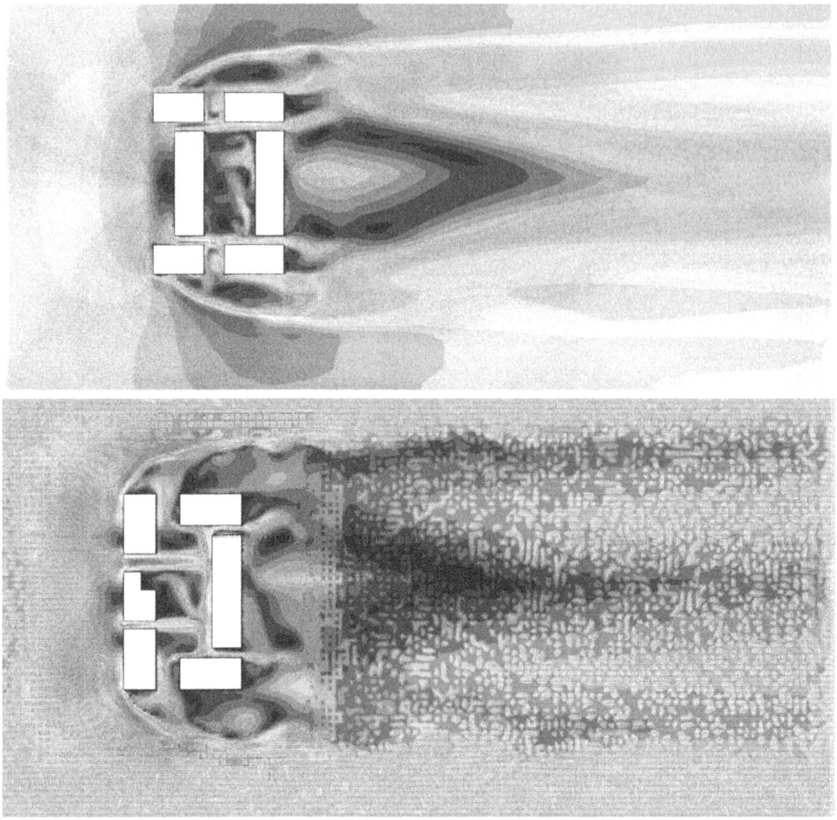

Figure 14.15 Douala Cultural Hub, Fluid Dynamic Analysis.

Figure 14.16 Douala Cultural Hub, © Dione Roach.

Figure 14.17 Douala Cultural Hub, Cameroon © Thrésor Michel Ndasse.

Figure 14.18 Douala Cultural Hub, Cameroon © COE.

Figure 14.19 Douala Cultural Hub, Cameroon © Dione Roach.

Figure 14.20 Douala Cultural Hub, Cameroon © Dione Roach.

Bibliography

Lepik, A., Beygo, A., 2016, *Francis Kéré: Radically Simple*, Hatje Cantz Verlag.
Eglash, R., 1999. *African Fractals*, Rutgers University Press.
Cinqualbre, O., 2009. Jean Prouve, The Tropical House. Centre Georges Pompidou Service.
Beguin, J.P., Kalt, M., Leroy, J.L., Louis, D., Macary, J, Pelloux, P., Peronne H.N. 1952. L'habitat du Cameroun. Paris: publication de l'office de la recherché scientifique outre-mer; edition de l'union francaise.
Cascone P., Laddaga M., 2019. African Fabbers House Digital Manufacturing for Sustainable Living, *Responsive Cities: Disrupting Through Circular Design Symposium Proceedings 2019*, ISBN-13 978-84-120885-1-9.
Filosa F., 2022, Douala Cultural Hub, *Platform 33*, ISSN 2420-9090.

Machine Not Homed

Growth and Perspectives on Digital Fabrication Made in Latin America

Pablo C. Herrera

Introduction: Global North and the Global South Context

To write the history of digital fabrication requires revisiting some key thinkers. In the Latin version of the first book of Vitruvius, for example, the word "craftsmanship" appears as a translation of the Latin "Fabrica", and "technology" as the translation of the Latin "ratiocinatione".[1] For Vitruvius, "fabrica (craft) is continued and familiar practice, which is carried out by the hands".[2] Vilem, stated that the history of fabrication (*fabrikation*) is the history of societies and industrial revolutions, and that "fabrication means the same thing as learning".[3] Flusser also states that in the future the "factory" will become a "school" and suggests that the history of fabrication is a transformation of four phases: Appropriation (*entwenden*) or hand-made, Conversion (*umwenden*) or tool-man, Application (*unwenden*) or machine-man, and Use (*verwenden*) or robot-man. Similarly, Lewis Mumford argued that the evolution of techniques is the history of the hand-made, tools (manipulation), machines (automatic action), and machine tools. In the recent context of fabrication laboratories, Neil Gershenfeld characterized the relationships among analog computers, mainframes, personal computers, tools machines, and personal fabricators as a transition toward precision regardless of its complexity.[4]

The globalization of digital fabrication began in 2001 and coincided with the emergence of Do-It-Yourself (DIY) and Do It With Others around 2001. The various interactions in the contexts of manufacturing, DIY culture, instrumentalization, and rapid

 DOI: 10.4324/9781003296522-18

prototyping determined the cultural evolution of digital fabrication in architecture in the West, and Latin America was no exception. The purpose of this essay is to discuss the origins and trajectories of digital fabrication in Latin America. Here, we map the evolution of the field in 20 Latin American countries, during the first 20 years of the 21st century. The focus is the analysis of resources of academic production, their different levels of scientific publication and their interaction with practice around the construction industry ecosystem and fabrication laboratories. This essay sheds light on digital fabrication in countries such as Argentina, Brazil, Chile, Colombia, Mexico, Peru, Uruguay, and a few in Central America. This history provides an alternative lens to reconstruct the lineage of digital fabrication through an extensive mapping and analysis of projects, the role of the government, academics, universities, conferences, and research groups, directly or indirectly linked to different Latin Americans who implemented digital fabrication in the region (Figure 15.1).

Latin America and the Route from Knowing to Making

The origin of architectural digital fabrication is a history of knowing (theory) and making (practice). The milestones and interactions of this history in the Global North reveal how manufacturing processes were incorporated into the practice of architecture and pedagogy in academia, which gave rise to the culture of personal fabrication. The literature on digital fabrication shows an evolution of topics, machines, techniques, flows, processes, and objects on different scales. Scholars such as Antoine Picon have argued that the technological transformations of the 20th century are inseparable from globalization; they occurred through long complex processes, beyond the history and implementation of digital tools.[5] Similarly, George Basalla has argued that technological transformation is often configured by economic, military, political, and cultural events.[6] After the Second World War, for example, the history of computers and North American manufacturing is one of alliances among companies, universities, and the government.[7] Scholars such as Daniel Cardoso Llach have analyzed this convergence of the military, industry, and the academy with the first numerically controlled milling machines and the first CAD technologies in the 1950s.[8]

In the Global North, therefore, instrumentalization promoted digital fabrication, originating in practice and the cultural movements associated with digital tools such as CAD and Computer aided manufacturing (CAM). Alternatively, in South America, knowledge spread from initiatives by Latin American academics who traveled to conferences in the North or completed doctorates or visits to the northern hemisphere and who experimented with rapid prototyping and fabrication techniques in architecture outside of Latin America. Contrary to the evolution of digital fabrication in the Global North, digital fabrication in Latin America evolved not through industry or manufacturing, but in the context of theoretical knowledge (knowing) first and then practical knowledge (making).

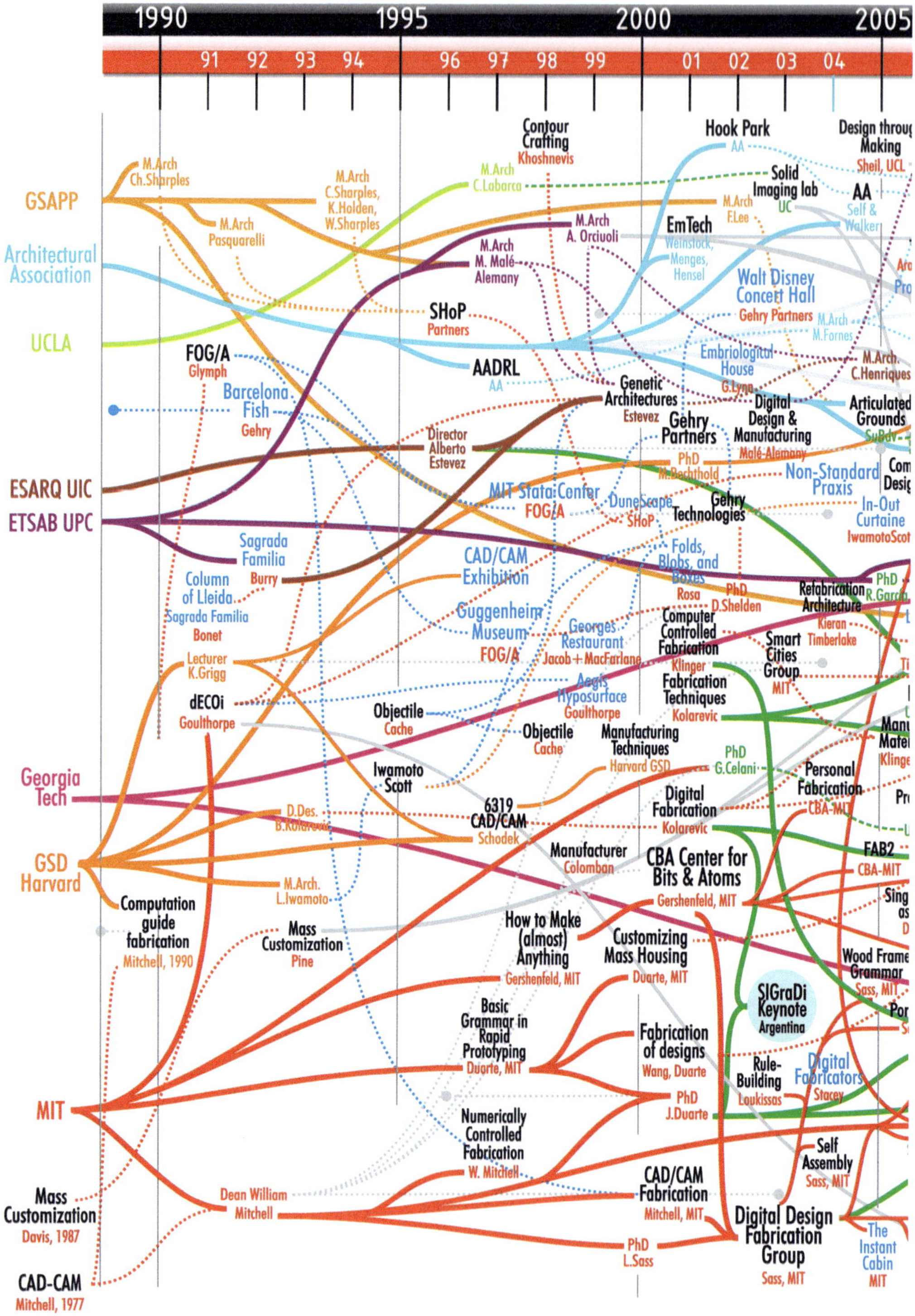

Figure 15.1 Mapping Digital Fabrication between the Global North and Global South. 1990–2021.

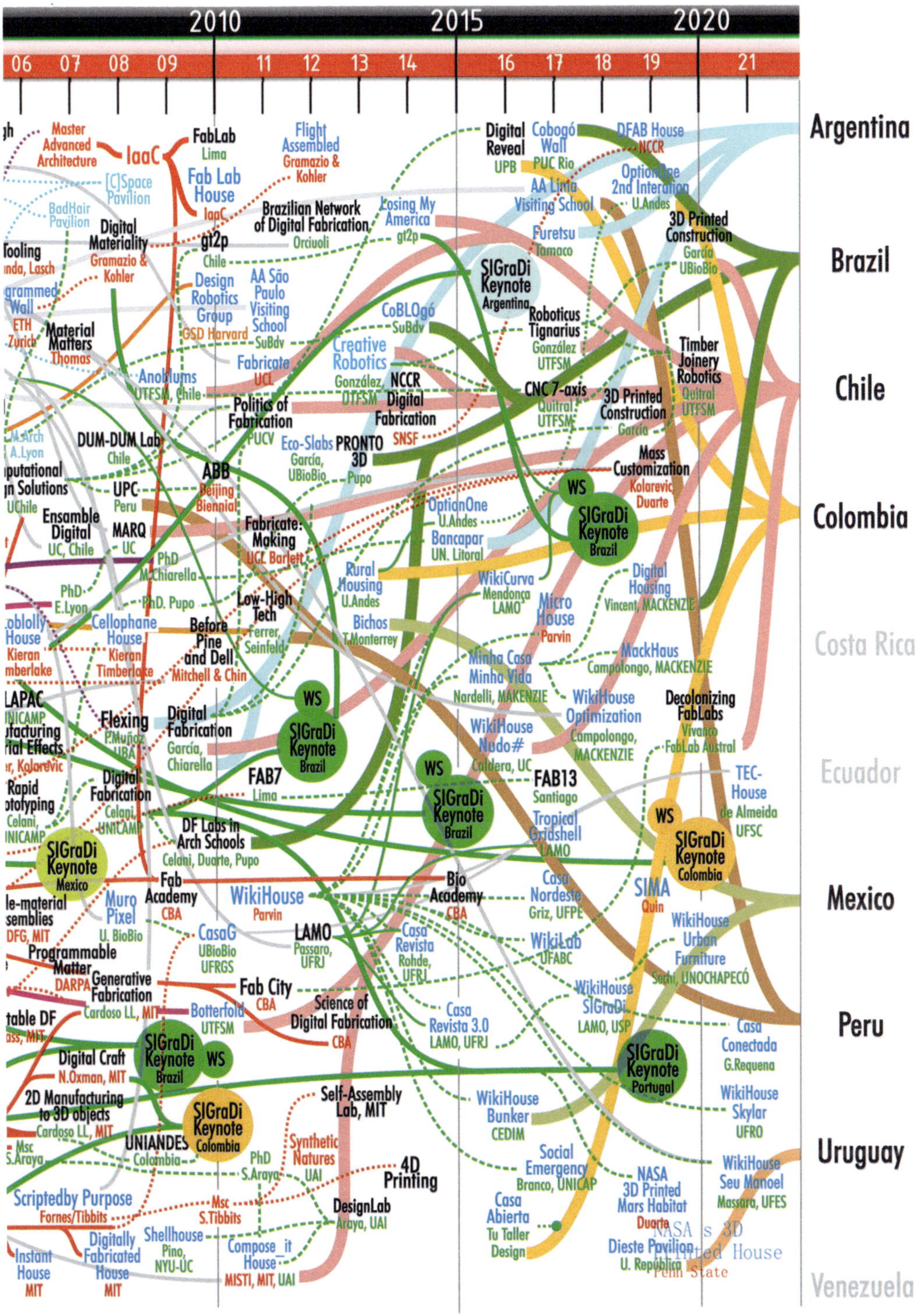
2010
2015
2020
06 07 08 09 11 12 13 14 16 17 18 19 21
Argentina
Brazil
Chile
Colombia
Costa Rica
Ecuador
Mexico
Peru
Uruguay
Venezuela
Master Advanced Architecture
IaaC
FabLab Lima
Fab Lab House IaaC
[C]Space Pavilion
BadHair Pavilion
Digital Materiality Gramazio & Kohler
gt2p Chile
Flight Assembled Gramazio & Kohler
Brazilian Network of Digital Fabrication Orciuoli
Losing My America gt2p
Design Robotics Group GSD Harvard
AA São Paulo Visiting School SuBdv
Material Matters Thomas
Anoblums UTFSM, Chile
Fabricate UCL
CoBLOgó SuBdv
Creative Robotics González, UTFSM
NCCR Digital Fabrication SNSF
Politics of Fabrication PUCV
Eco-Slabs García, UBioBio
PRONTO 3D Pupo
DUM-DUM Lab Chile
UPC Peru
MARQ UC
Ensamble Digital UC, Chile
ABB Beijing Biennial
Fabricate: Making UCL Barlett
PhD M.Chiarella
PhD E.Lyon
PhD. Pupo
Low-High Tech Ferrer, Seinfeld
Cellophane House Kieran Timberlake
Before Pine and Dell Mitchell & Chin
Digital Reveal UPB
Cobogó Wall PUC Rio
AA Lima Visiting School
DFAB House NCCR
OptionOne 2nd Interation U.Andes
Furetsu Tamaco
SIGraDi Keynote Argentina
Roboticus Tignarius González UTFSM
3D Printed Construction García UBioBio
Timber Joinery Robotics Quitral UTFSM
CNC 7-axis Quitral UTFSM
3D Printed Construction García
Mass Customization Kolarevic, Duarte
WS
SIGraDi Keynote Brazil
OptionOne U.Andes
Bancapar UN. Litoral
Rural Housing U.Andes
Bichos T.Monterrey
WikiCurva Mendonça LAMO
Micro House Parvin
Digital Housing Vincent, MACKENZIE
MackHaus Campolongo, MACKENZIE
Minha Casa Minha Vida Nardelli, MAKENZIE
WikiHouse Optimization Campolongo, MACKENZIE
WikiHouse Nudo# Caldera, UC
Decolonizing FabLabs Vivanco FabLab Austral
Digital Fabrication García, Chiarella
Flexing P.Muñoz UBA
Digital Fabrication Celani, UNICAMP
FAB7 Lima
DF Labs in Arch Schools Celani, Duarte, Pupo
FAB13 Santiago
Tropical Gridshell LAMO
TEC-House de Almeida UFSC
SIGraDi Keynote Colombia
SIGraDi Keynote Mexico
Fab Academy CBA
Muro Pixel U. BioBio
WikiHouse Parvin
Bio Academy CBA
Casa Nordeste Griz, UFPE
SIMA Quin
CasaG UBioBio UFRGS
LAMO Passaro, UFRJ
Casa Revista Rohde, UFRJ
WikiLab UFABC
WikiHouse Urban Furniture Sechi, UNOCHAPECÓ
Programmable Matter DARPA
Generative Fabrication Cardoso LL, MIT
Fab City CBA
Botterfold UTFSM
Science of Digital Fabrication CBA
Casa Revista 3.0 LAMO, UFRJ
WikiHouse SIGraDi LAMO, USP
Casa Conectada G.Requena
Digital Craft N.Oxman, MIT
SIGraDi Keynote Portugal
2D Manufacturing to 3D objects Cardoso LL, MIT
Self-Assembly Lab, MIT
WikiHouse Bunker CEDIM
WikiHouse Skylar UFRO
UNIANDES Colombia
Synthetic Natures UAI
PhD S.Araya
Social Emergency Branco, UNICAP
NASA 3D Printed Mars Habitat Duarte
WikiHouse Seu Manoel Massara, UFES
4D Printing
Scriptedby Purpose Fornes/Tibbits
Msc S.Tibbits
DesignLab Araya, UAI
Casa Abierta Tu Taller Design
NASA's 3D Printed House Penn State
Dieste Pavilion U. Republica
Digitally Fabricated House MIT
Instant House MIT
Shellhouse Pino, NYU-UC
Compose_it House MISTI, MIT, UAI
LAPAC UNICAMP
Rapid Prototyping Celani, UNICAMP
Loblolly House Kieran Timberlake
Tooling
M.Arch A.Lyon
Computational Design Solutions UChile

From Knowing Culture

Knowing (theoretical knowledge) is the first context that evolved with computational strategies in workshops, schools, and university programs while it was continually being implemented in the northern hemisphere. In the South, there was no consolidated link with the manufacturing or construction industry. Our study maps the researchers, actors, and contexts that originated the foundation of laboratories in Latin American universities outside the Fab Foundation network, an international network of fabrication laboratories, to show how digital fabrication evolved in the South. In the first decade of the 21st century, experimentation focused on the personal interests of students, teachers, professionals, artisans, and government managers. In the second decade, our mapping of fabrication topics reveal projects aimed at local problems and needs in architecture schools, as well as in industry and communities.

Our research relies heavily on statistical data from academic conferences and papers archived in the Cumulative Index of Computer Aided Design database, an extensive bibliographic database initiated by Martens and Turk in 1999. We also draw from various other studies on keywords and subjects in Latin American technological conferences, workshops, projects, and events that illustrate the diversity of ideas and enormous spread of digital fabrication in the region.[9] For this essay, we added data from 2019 to 2020 to these previous studies to visualize the spread of digital fabrication knowledge.

Despite these academic efforts, our mapping of the region was only partially characterized and analyzed. In the missing countries, the instrumentalization or the degree of difference between the implementations of digital fabrication warrants further study. Nevertheless, we have been able to map the state of digital fabrication in 12 countries in South America, seven countries in Central America, and one country in North America from data produced over two decades by different researchers.

From Making Culture

Making (Practical Knowledge) is another aspect of digital fabrication that was strengthened by global DIY culture and the implementation initiatives of "Fab Labs" and maker culture. When this culture became globalized, its instrumentalization, assistance and advice, provided Latin America a suggestion for its implementation in universities.

The culture of Fab Labs in the Global North was exported in the first ten years of the 21st century with an academic initiative that linked Central America in an alliance with the "Instituto Tecnológico de Costa Rica". The Fab Labs concept began in the pedagogy section of the Media Lab's Physics and Media group at the Massachusetts Institute of Technology (MIT) with the creation of the MAS.863 How To Make (Almost) Anything in 1998. The National Science Foundation (NSF) funded the initiative between 2001 and 2007 with a total of US$14.2M awarded to the MIT Center for Bits and Atoms (CBA). The CBA-MIT consolidated the Fab Lab program, and Costa Rica was the only partner until 2008 when MAS.863 went global with the Fab Academy project and reached more countries.[10] In this program, 1,103 students graduated (until 2020), including 141 graduates from 23 Latin American Fab Labs. Since 2009, the Fab Foundation has managed the project and designated the first institution in the world to receive funding from the Spanish Agency for International Development Cooperation to implement a Fab Lab in South America.[11] The Faculty of Architecture, Urbanism, and Arts of the "Universidad Nacional de Ingeniería" in Peru, followed the guidelines of the Fab Lab program originating at CBA-MIT and the Fab Academy program at the Fab Lab Barcelona of the Instituto de Arquitectura Avanzada de Catalunya (IAAC) in Spain. Mapping these actors and initiatives allows us to see how the connections between the North and the South evolved.

(Re)growing Across Inequalities

It is important to note that Latin America goes through positive cycles for technological implementations and negative periods that cause delays and distancing from global progress. Military dictatorships have been a constant in this process. For example, technological initiatives were dismantled in 1973 by the military dictatorship in Chile. Research such as "automatic drawing" by Arturo Montagú, founder of the Sociedad Iberoamericana de Gráfica Digital (SIGraDi), continued between the Architectural Association and the University of Manchester in 1967–1968 as a consequence of the coup d'état in Argentina in 1966.[12] Dictatorships continued and Latin America's problems worsened in the "lost decade" of the 1980s, prolonging the increase in its debt until 1990. Other realities include high rates of inequality and adverse conditions in professional and academic worlds (low cultural esteem, few full-time laboratory positions, limited financial resources, etc.). These are latent features throughout the Global South that characterize the region's relationship with emerging technologies.

In 2011, different efforts culminated in Peru, including the installation of the first laboratory certified by the Fab Foundation in South America. Its inauguration coincided with the World Fablabs (FAB7) meeting. In our study of this event, we concluded that ideas related to Multiculturalism, Eco-diversity, and Social Capital were highlighted as opportunities to expand the use of technology and move the economic and social development of communities.[13]

At this time, Latin American architecture was growing in the global shadow of a radical and experimental architecture focused on variations in form and its transformation, derived from rules, relationships, and generative principles combining genetics and algorithmic processes. It took some time for these ideas to reach the Global South.

Except for the projects of the Chilean Mathías Klotz produced between 1995 and 2000 for Archilab 2000 and the participation of Latin Americans as postgraduate students of the Architectural Association Design Research Laboratory (AADRL) in the Far Eastern International Digital Architecture Design (FEIDAD) exhibitions, there is no evidence of direct participation of Latin America in global exhibitions until 2008. The first Latin American digital fabrication experiments were curated by Matias del Campo for the Beijing Architecture Biennial. Projects produced in Brazil, Colombia, Mexico, and Peru presented examples of computational design experimentations in 2008 and design

Figure 15.2 CoBLOgó. Factory & Office Annex. São Paulo, Brazil, 2014. Credits: SUBdv.

computing linked to machinic processes in 2010. However, after 2010, there was little Latin American presence in an international context until 2014 at the exhibition "New Territories. Laboratories for Design, Craft, and Art in Latin America" with projects from the offices Great Things to People (gt2p) from Chile and Guto Requena from Brazil.[14]

The presence of local experience with digital fabrication globally returned in 2015, with regional projects in the exhibition "Homo Faber: Digital Fabrication in Latin America", for the CAAD Futures 2015 world conference, held in São Paulo, Brazil. Figure 15.2 shows one of the 24 exhibited projects by the office SUBdv from Brazil.

Between "Machine not Homed" and "Homing the Machine"

In the Global North, industry and academia are the homes referenced by the operational state of "Homing the Machine". However, in a territory like Latin America "that is often underrepresented, excluded, or ignored in the global narrative (...) shaped by long histories of imperialism, colonization, and global asymmetry" the status "Machine not homed" more accurately describes the early years of the 21st century.[15]

Latin America has advanced computationally, but often without machines that materialize ideas. To understand the origin of digital fabrication in Latin America, we require a comparative analysis of foundational motivations and implementations different from the Global North. Although these differences diminished after the first decade of the 21st century, they are still commonplace in those parts of Latin America that have not yet overcome the implementation difficulties. Table 15.1 identifies inequities of context and specificity between the Global North and Global South in the realm of digital fabrication.

Table 15.1 Realities of Digital Fabrication between the Global North & Global South

		Global North	Global South
[A]	Economic Participation	Economically developed, high-income, politically stable, and culturally dominant	Economically dependent, politically unstable, and at the mercy of cultural trends
[B]	Industry	Global manufacturing, transfer technology, Tech inventive	informal, imitation-based, low-cost labour, colonial experience
[C]	Funds/ Grants Source	Military, Government, Industry	Academy, Private Resources
[D]	Initial Lab Size	High-End Lab	Prototyping Facility
[E]	Lab Size (Average)	Very High-End Lab	Fab Lab
[F]	Project Scales	Architectural Scale (1:1 construction)	Objectual (scaled models)
[G]	Market Demand	Commercial, Industry, Academy	Academy
[H]	Discipline participation	Cross-Disciplinary Ways	Architects or Engineers
[I]	First impact of Scripting Culture	Digital Fabrication (practice/academy)	Design (academy)
[J]	Computational Evolution	1°Digital Fabrication, 2°Design	1°Design, 2°Digital Fabrication
[K]	Audience	Upper-Class Academics, hobbyists: well-educated but impractical students	Entrepreneurs, graduate students, artisans, teachers, interested in innovation, middle, and lower class

One of the biggest discrepancies is economic support. Whereas the implementation of digital fabrication in the northern hemisphere began with investments by government agencies, especially the NSF, this type of support was not very common in Latin America.[16] However, Brazil and Chile did use the competitive financing models of their countries to develop Fab Labs. In Chile, incentive programs first focused on innovative methodologies in academic programs. Three programs encouraged the implementation of infrastructure: the Higher Education Quality Improvement Program (MECESUP), The National Fund for Scientific and Technological Development of Chile (FONDECYT), and Fondo de Innovación para la Competitividad (FIC). Universities and architecture studios linked their initiatives to these financings. The MECESUP was an alliance between the World Bank and the Government of Chile, which financed initiatives related to digital fabrication in architecture schools of the Pontificia Universidad Católica in 2003, Universidad de Chile in 2008 and 2012, Universidad Técnica Federico Santa María in 2009, and Adolfo Ibáñez University in 2013.

Two notable examples of funded labs came from Chile and Brazil. Gt2p is a Chilean architecture and design studio founded in 2009. Its projects range from digital crafting to low-tech knowledge. They received two rounds of seed capital from the Technological Innovation Fund of Chile for the parametric design and digital manufacturing of furniture. In Campinas, Brazil, the "Laboratório de Automação e Prototipagem para Arquitetura e Construção" (LAPAC) was created by Gabriela Celani at the *Universidade Estadual de Campinas* (UNICAMP) in 2007. With their experience and the vision of the UNICAMP authorities, they inaugurated the "Espaço Plasma" in 2020 to serve the entire academic community.

In Santiago de Chile, the recently inaugurated "Robotic Construction and Manufacturing Lab" emerged from the "Creative Robotics" (2013) and the "Laboratorio de Materialidad y Fabricación Digital" or LABOMAT (2008) promoted by Luis Felipe González and Francisco Quitral at the "Federico Santa María Technical University" (UTFSM) (Figure 15.3).

In Concepción, Chile, the "Pilot Scale Physical Prototyping Laboratory (PEP) (Figure 15.4)", located at *Universidad del Bio-Bio* (UBB), became part of an implementation network within the framework of the Construye2025 investment project in Chile. This implementation is the result of applications for financing by teams led by Rodrigo García Alvarado since 2008 and which, at the same time, promoted synergy with academics in digital fabrication between Brazil and Argentina.[17]

These financing models are still being introduced in some countries of the Andean region, but in others the situation does not have a defined horizon. For now,

Figure 15.3 Robotic Construction and Manufacturing Lab, Valparaiso, Chile. Photo: F. Quitral.

Figure 15.4 Pilot Scale Physical Prototyping Laboratory, Concepción, Chile. Photo: R. García.

investments are private, from universities that are certain that technology has value in the professional future of the architect and that this training will allow them to enter a global market. What the above examples illustrate is how the machine might start to feel at home in Latin America.

Self-learning Without Borders

The generation of Latin Americans who returned from postgraduate studies in the late 1990s and early 21st century was key to the processes of institutionalization of technologies in Latin America. However, toward the end of the first decade of the 21st century, the implementation of design computing and digital fabrication changed radically in the Global South and developed patterns different from the traditional forms of acquisition through postgraduate studies in the Global North.

The financial crisis of 2008 was one of the determining factors for migration between different geographies. Universities from the northern hemisphere coordinated visits to the south, and Latin American academics with offices and courses in the north returned and shared their experience in their home countries. This exchange and mobility marked a new way of learning and empowering the local and global in both directions.

Academic knowledge also spread through visits and global travel by experts in digital fabrication. For example, doctoral and master's students from MIT's Digital Design Fabrication Group conducted workshops in Chile in 2006–2007 and Peru in 2008. Before inaugurating the Self-Assembly Lab at MIT, Skylar Tibbits co-directed a workshop in Colombia in 2009 with Daniel Cardoso Llach. Academic programs such as the AADRL enrolled 60 Latin American students between 1997 and 2007; this grew into the AA Visiting School in 2008, which in a decade made more than a hundred visits to Latin America. IAAC was founded in Spain (2003), and until 2008 received 46 Latin Americans before launching the Fab Academy together with the MIT Center for Bits & Atoms in 2009, the world's first academic training program in digital fabrication with 1,379 enrollments between 2009 and 2021 and 10% from Central and South America. Instructors such as Affonso Orciuoli from the Bio-Architecture program of the Universitat International de Catalunya (UIC) gave workshops in Peru in 2006, in Colombia in 2010, and in Brazil in 2009 and 2011. Orciuoli established the Brazilian Network of Digital Fabrication in 2012 and joined different digital fabrication research groups.

However, temporary visits and workshops were not the only instances that changed learning. The ubiquity and transmission of knowledge by code allowed reuse of the techniques produced. In the northern hemisphere, self-learning was closely tied to manufacturing and continued through networked communities.[18] Before Massive Open Online Courses, software developers, master's, and doctoral students in architecture, and instructors with workshops and courses used Blogger, WordPress, and Wikis, to share code and foster discussion spaces. Between 2003 and 2009, instrumentalization knowledge and digital fabrication techniques spread globally online. An emerging resource production ecosystem formed in Europe (46 blogs), the United States (26 blogs), and Latin America (13 blogs) with more than 300 resources and codes available to all and were consumed by universities and offices around the world.

At this time, computational and fabrication requirements in construction became more intense, inspiring 12 offices around the world to form computational groups to solve complex situations. These positions were attractive to Latin Americans who finished their studies in the northern hemisphere. Some of the offices that employed them include ARUP (Advanced Geometry Unit), Buro Happold (Smart Space), Foster + Partners Studio (Specialist Modeling Group, SMG), Gehry Partners (Gehry Technologies), Grimshaw (Grimshaw Design Technology Group), Herzog & de Meuron (Digital Technology Group), Skidmore, Owings & Merrill (Black Box Studio), UN Studio (SPP), and Zaha Hadid Architects (ZHA_CODE).

This type of alternative learning without borders modified (but did not eliminate) the traditional pipeline to postgraduate studies in emerging technologies that kept southern migrants in the northern hemisphere for two or five years.

Digital Fabrication Challenges and Perspectives

The Latin American digital fabrication route was not the fortuitous or casual result of the evolution of the Global North experience. In our study, we found factors that accelerated implementation, beyond just the traditional experience of returning master's or doctoral students to their countries of origin. Of course, movements from the South to the cities of production of technology in the North facilitated the acquisition of new techniques. However, we found that while these trips influenced the consolidation of the infrastructure, they were just the starting point of knowledge sharing between the Global North and the South. The events and data analyzed suggest that the end of the first decade of the 21st-century global events affected and accelerated the implementation of technologies, and digital fabrication was no exception. The introduction of websites and online

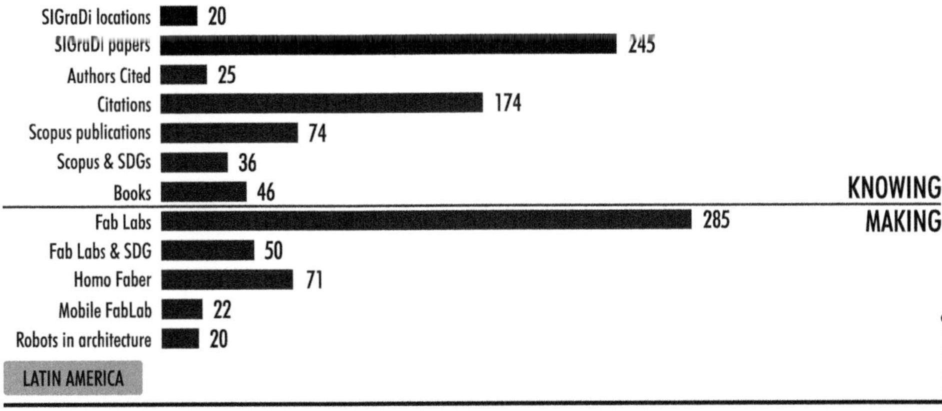

Figure 15.5 Panorama of digital fabrication in Latin America by numbers.

forums provided vast amounts of skills and code that could be consumed easily. This access radically accelerated the acquisition not only of information but also of knowledge. People became empowered with the exposure and reuse of code, self-learning, and the growth of Fab Labs and maker culture. Architects were the first link to the on-premises implementation and remain so to this day. They promoted in Latin America a culture of knowing, first the know-what and then the know-how.

University implementations happened with advice from their academics and other associates of the Fab Foundation. Today, governments encourage alliances between universities and industry, including other governments. This has given universities the role of ensuring an infrastructure that provides a service to society and its students. In Figure 15.5, we see a broad picture of the most recent state of digital fabrication knowing and making in Latin America.

While this data indicates positive growth, we can also see that the participation of Latin American countries is unequal. The indicators not only reveal the maturity of the topic but also warn about technological and gender gaps, mainly because the opportunities for the socialization of technology did not reach all the countries of the Andean region nor of Central America or the Caribbean. In this last section, we present a summarized vision of the three aspects that define technological, gender, and sustainability gaps and opportunities for further study.

Technological Gap

There are many countries in Central America and the Caribbean that are lacking representation in this field. Figure 15.6 shows the participation of countries using the global indicators summarized in Figure 15.5. This image succinctly illustrates an overview of digital fabrication in Latin American countries and the Caribbean.

SIGraDi is an important context in these findings. In 2019, Sperling et al. showed how digital fabrication has been a constant in the top five ranked keywords (Figure 15.7). Organizing a SIGraDi conference often showcases the advances of the region, but of course cannot necessarily intervene in the technological advancement of

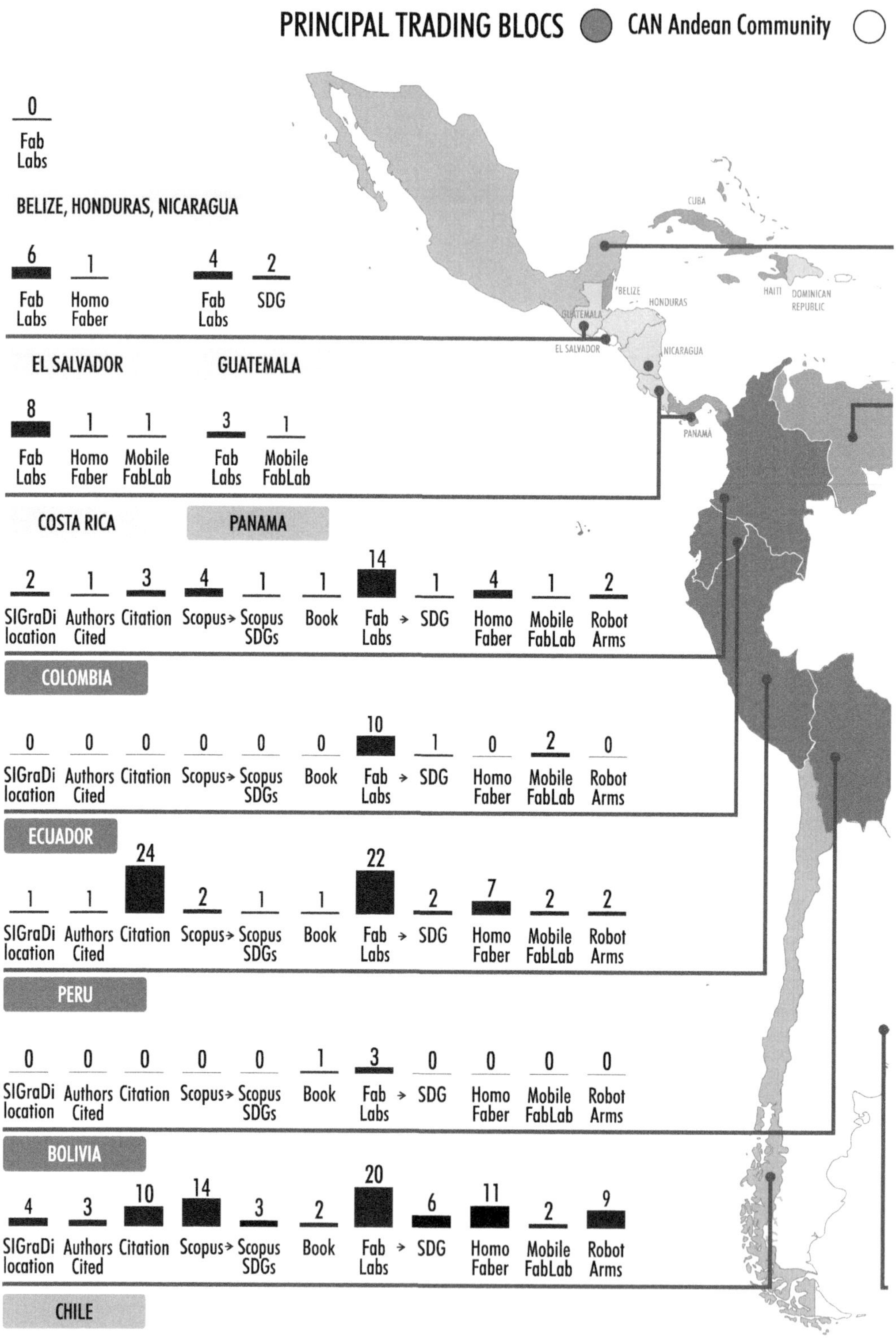

Figure 15.6 Overview of Digital Fabrication in Latin American countries and the Caribbean.

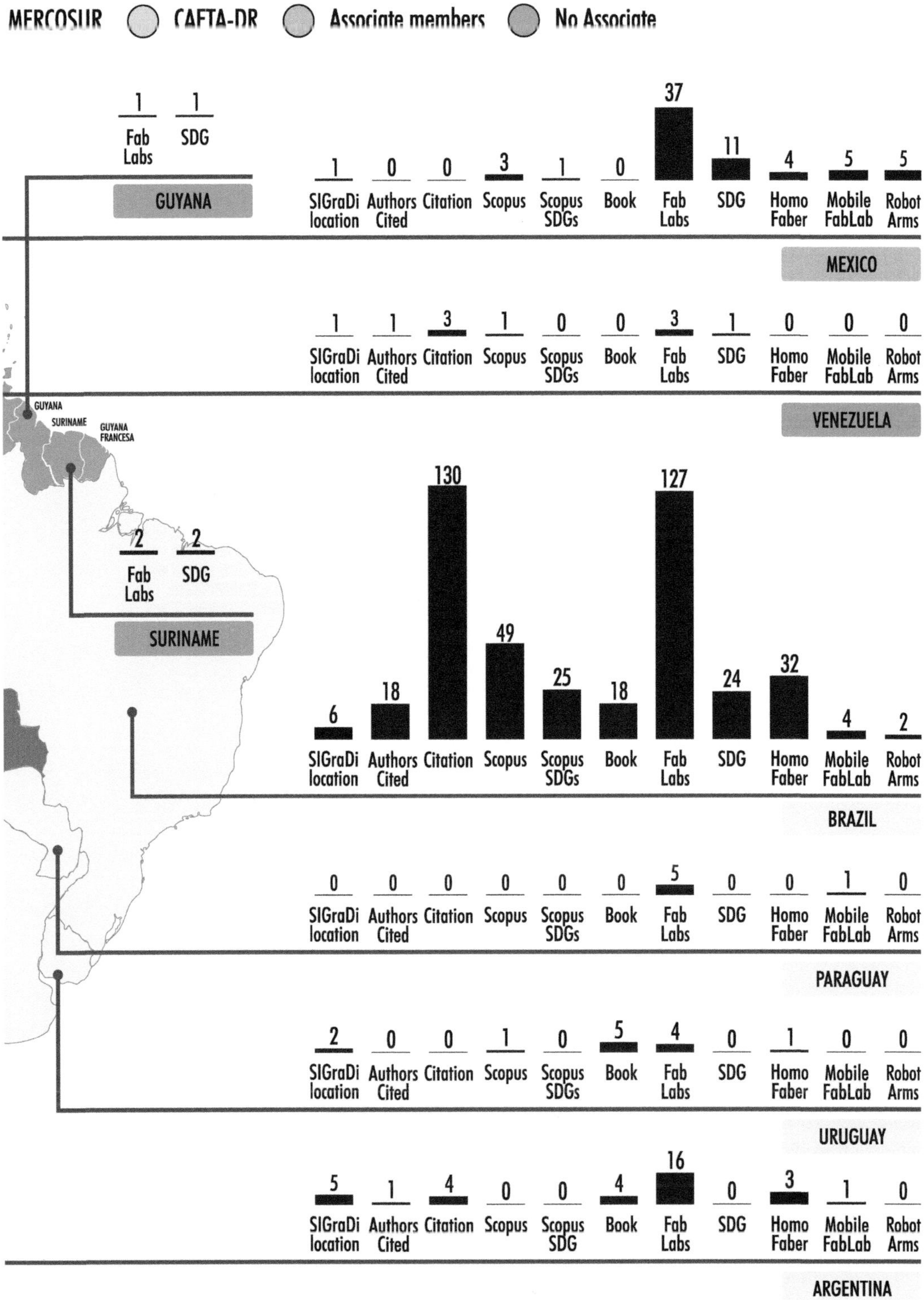
MERCOSUR
CAFTA-DR
Associate members
No Associate
1 Fab Labs
1 SDG
GUYANA
SIGraDi location 1
Authors Cited 0
Citation 0
Scopus 3
Scopus SDGs 1
Book 0
Fab Labs 37
SDG 11
Homo Faber 4
Mobile FabLab 5
Robot Arms 5
MEXICO
SIGraDi location 1
Authors Cited 1
Citation 3
Scopus 1
Scopus SDGs 0
Book 0
Fab Labs 3
SDG 1
Homo Faber 0
Mobile FabLab 0
Robot Arms 0
VENEZUELA
GUYANA
SURINAME
GUYANA FRANCESA
2 Fab Labs
2 SDG
SURINAME
SIGraDi location 6
Authors Cited 18
Citation 130
Scopus 49
Scopus SDGs 25
Book 18
Fab Labs 127
SDG 24
Homo Faber 32
Mobile FabLab 4
Robot Arms 2
BRAZIL
SIGraDi location 0
Authors Cited 0
Citation 0
Scopus 0
Scopus SDGs 0
Book 0
Fab Labs 5
SDG 0
Homo Faber 0
Mobile FabLab 1
Robot Arms 0
PARAGUAY
SIGraDi location 2
Authors Cited 0
Citation 0
Scopus 1
Scopus SDGs 0
Book 5
Fab Labs 4
SDG 0
Homo Faber 1
Mobile FabLab 0
Robot Arms 0
URUGUAY
SIGraDi location 5
Authors Cited 1
Citation 4
Scopus 0
Scopus SDG 0
Book 4
Fab Labs 16
SDG 0
Homo Faber 3
Mobile FabLab 1
Robot Arms 0
ARGENTINA

Figure 15.7 Digital Fabrication in the Top five rank keywords of the SIGraDi conferences (2006–2020).

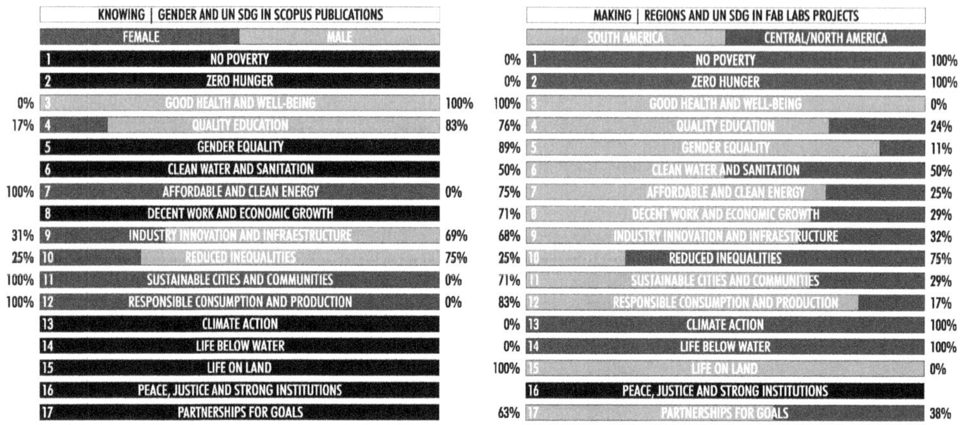

Figure 15.8 Digital Fabrication in Latin America and UN SDGs.

an entire country. Conference data allows us to see the number of articles on digital fabrication coming from these countries, but on the other hand it does not give clear reasons why the topic has not spread evenly. We must also note that government research policy affects indexed publications such as Scopus publications. In Brazil, Chile, Colombia, and Peru, it is a national policy that incentives are granted to articles indexed in Scopus or Web of Science, and this indicator does not apply to Argentina, which does not have this limitation. Therefore, publications in certain media depend on research funding policies. This factor encouraged us to gather other types of non-indexed publications, such as independent books.

Sustainability Gap

In 2015, the United Nations established 17 Sustainable Development Goals (SDGs) that guide global initiatives. These SDGs often guide technological implementation in Latin American countries. For our study, we identified SDGs related to the integration of Fab Labs and related technology in different countries (Figure 15.8). Scopus Publication topics and keywords reveal that different countries align themselves with different goals. Brazil, Chile, Colombia, Mexico, Peru, and Venezuela, for example, present publications aligned to seven different SDGs.

The stated interests and goals of Fab Labs also vary by region. In Figure 15.8, we show the different interests of South American Fab Labs compared to those in Central/North America. In the Center and North, SDG1 (No Poverty) and SDG1 (Zero Hunger) are targets of laboratories in both sub regions. In the South, SDG3 (Good Health and Well-Being) and SDG15 (Life on Land) were highlighted by laboratories in the region. Naturally, the breadth of the territory creates agendas with different interests. However, the challenge for the region is to empathize with the reality of the countries, capitalizing on experiences, and continuing with shared knowledge alliances.

Gender Gap

In 2020, we started a project about the presence of women in fabrication laboratories.[19] This project mapped the total number of students who participated in Fab Academy between 2008 and 2021. The research focused on Fab Lab-based learning as an inclusive methodology. It revealed that the percentage of participation of women is around 30%, higher than the average 20% presence of women in other Science, technology, engineering, art, math (STEAM) and engineering environments. Nevertheless, it is still an unequal field in terms of gender (Figure 15.9).

We have a significant challenge in this aspect, mainly because Fab Lab-based learning focuses on solving problems without thinking about gender. That being said, Fab Labs have been culturally male-dominated. Questions of inclusion perhaps cannot be answered by technology but instead by the culture.

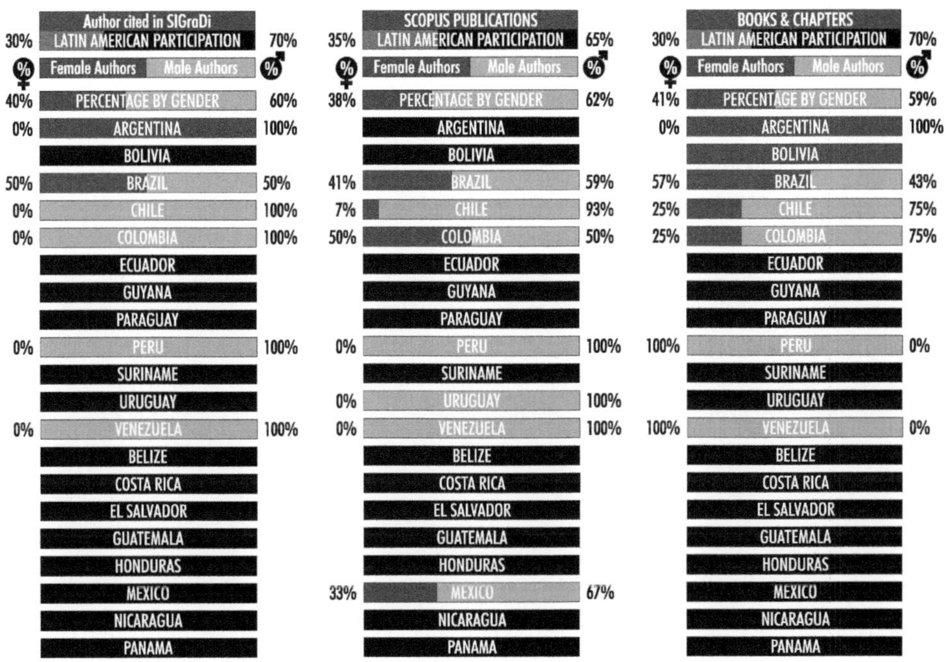

Figure 15.9 Country participation and Gender inside digital fabrication in Latin America.

A cultural shift can be seen in the local blending of making and knowing that has matured into concepts and editorial agendas that bring together visions from different regions of Latin America. Ideas and themes such as "New Territories", "Tropicalized digital aesthetic", "Paracrafting" and "Manufactured Landscapes", "Other Computations", "Computation Regionalism", and "Digitalities of the South" are concepts that have captured the potential of digital fabrication and its interaction with the locality, its materials, its community, landscape, city, and housing.[20] This expansion of topics may lead to more inclusion in the field.

Conclusion

What we have presented here is a comprehensive picture of the development of digital fabrication in Latin America. It is our hope that the complex relationships and interactions outlined here demonstrate that this development is more than just the exportation of knowledge from the Global North. Of course, when generalizing from observation, interpretation, and analysis of the panorama of digital fabrication in Latin America, omissions are inevitable. However, our goal with this research was to gather the largest amount of data produced by the 20 countries of the region in 20 years. We hope that the analyzed interactions between people, technologies, and institutions provide the starting point to build more in-depth investigations of material culture in each country and in the broader context of digital fabrication.

Appendix: Mapping Digital Fabrication in Latin America

Here, we present a statistical summary of our comprehensive study. As previously mentioned, this research was focused on the interaction of knowing and making. In addition to the personal histories of people and institutions discussed above, we have produced a set of diagrams that allow us to visualize the interaction flows and correlations that drove digital fabrication in Latin America between 2000 and 2020 (Figure 15.10). These diagrams deconstruct digital fabrication and production in countries, institutions, and fabrication laboratories, which characterize the concentration of certain knowledge in specific regions. The results not only allow us to understand the evolution of digital fabrication in the countries that promoted it but also illustrate the effects of academic migration, institutional outreach, and global platforms on technology in the Global South (Figure 15.11).

Mapping Knowing Culture in Digital Fabrication

These diagrams track the intellectual journey of digital fabrication discourse. In Figure 15.12, for example, we see the mapping of books produced in the northern hemisphere between 2000 and 2020. Here, we see the role of curators and conference chairs on the development of digital fabrication discourse. It is characterized by the panorama of

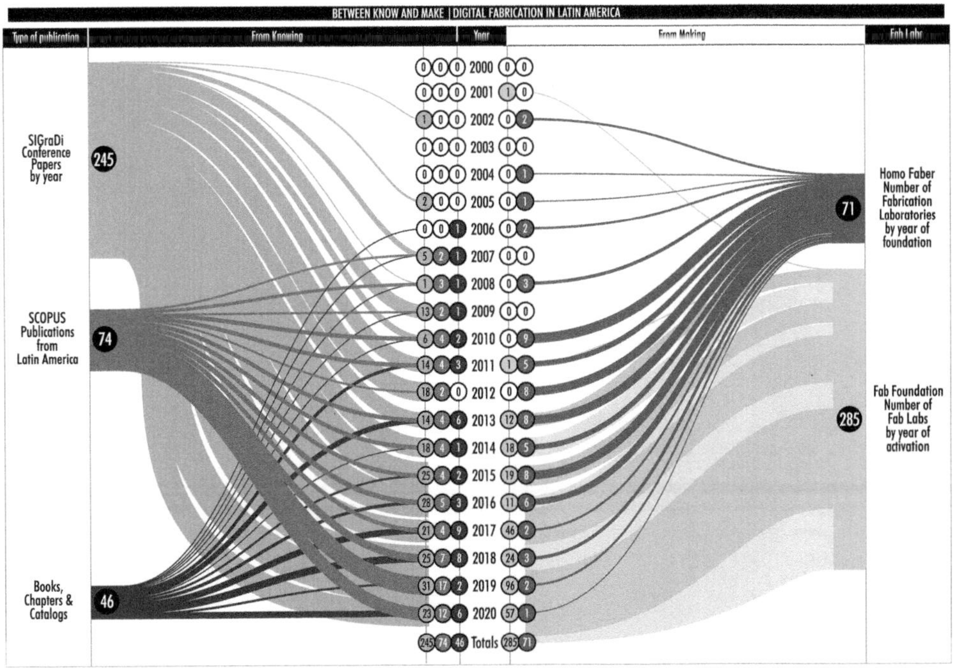

Figure 15.10 2000–2020. Between knowing (number of publications) and making (number of fab labs).

non-standard architecture and supported by manufacturing. Publications followed the organization of techniques and categorization of its instruments. The publications included experiments in the academy as well as in practice and showed the evolution of the different aspects of personalized and assertive fabrication that made possible the materialization and dissemination of ideas.

In Latin America, the SIGraDi has been promoting the exchange and discussion of ideas between architects and designers since 1998, mainly from the academy. Its establishment has motivated a series of bibliometric studies including our own. In Figure 15.13, we visualize in green the percentage of articles that directly reference digital fabrication as a topic, whereas Figure 15.14 tracks citations by country. Figures 15.15 and 15.16 expand on publication data and show how knowledge was distributed through conference papers, articles, and book chapters.

Since 2008, books and catalogs have also reflected the interests of the Latin American academic community in digital fabrication. This media is direct evidence from workshops, exhibitions, and practical explorations that emerged from the implementation of digital fabrication laboratories during the first decade of the 21st century. Figure 15.17 shows data from 11 online library catalogs, selected from the Top Eleven Rank Latin American Universities produced by the SCIMAGO Institutions Rankings (SCIMAGO, 2023). We included repositories from Brazil, Chile, Colombia, Mexico, Argentina, Uruguay, and Peru, which gave us a total of 46 items with records of "fabricación digital" (Spanish) and "fabricação digital" (Portuguese) in books, chapters, dossiers, and catalogs.

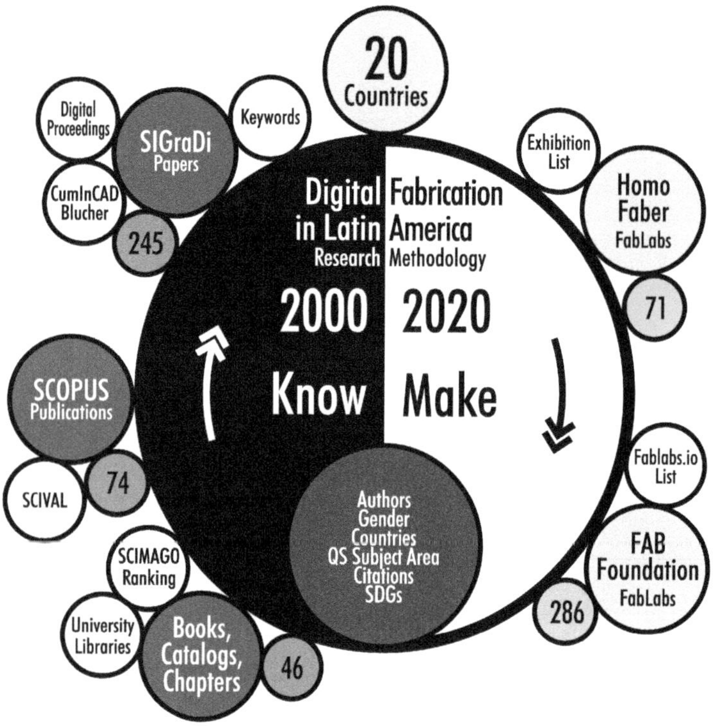

Figure 15.11 Methodology. Universe of knowing and making.

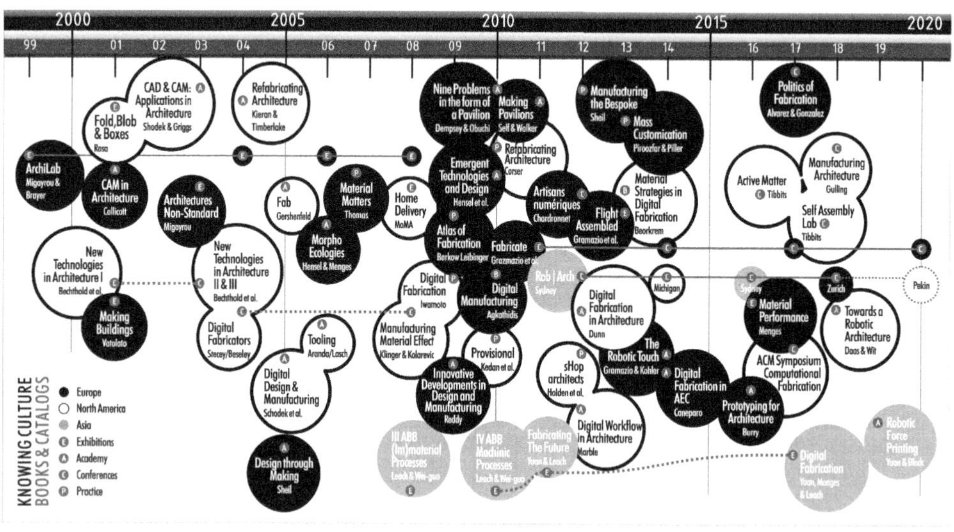

Figure 15.12 Digital Fabrication. Global mapping of books and catalogs 2000–2020.

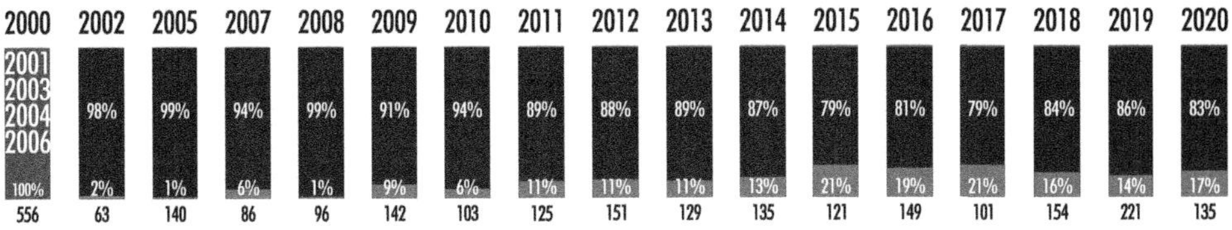

Figure 15.13 Digital Fabrication Conference Papers in SIGraDi 2000–2020.

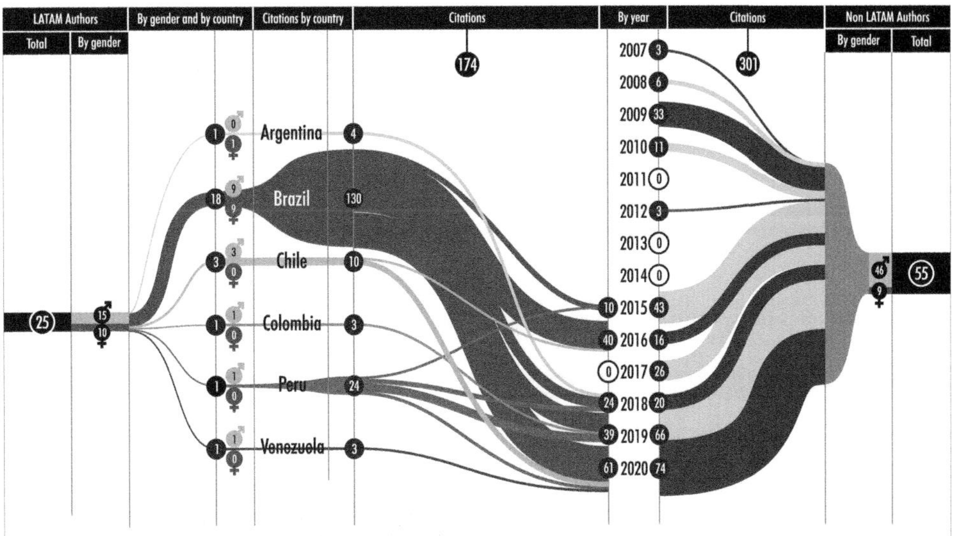

Figure 15.14 Digital Fabrication authors and citations in SIGraDi conferences.

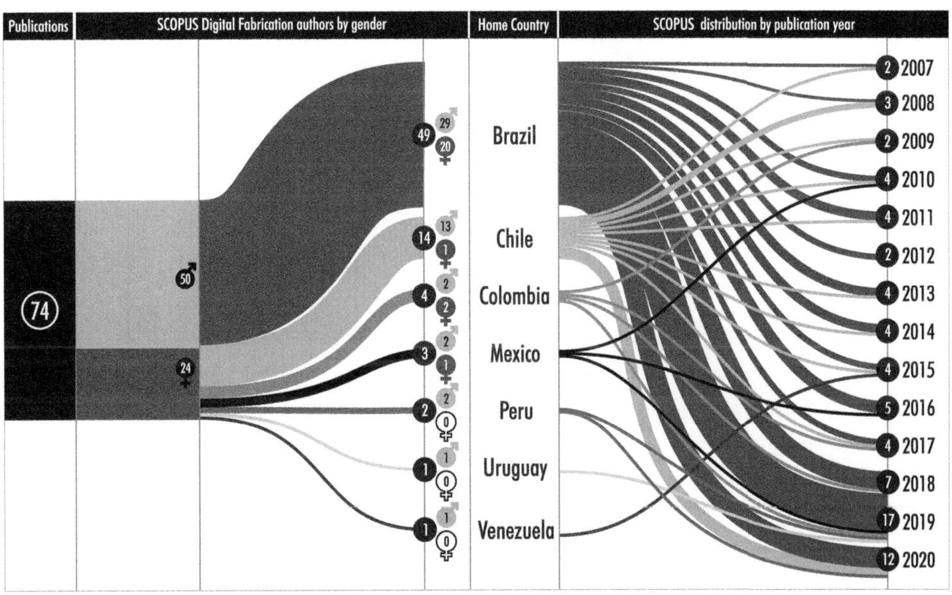

Figure 15.15 Scopus publications about Digital Fabrication. First Author genre and distribution by year.

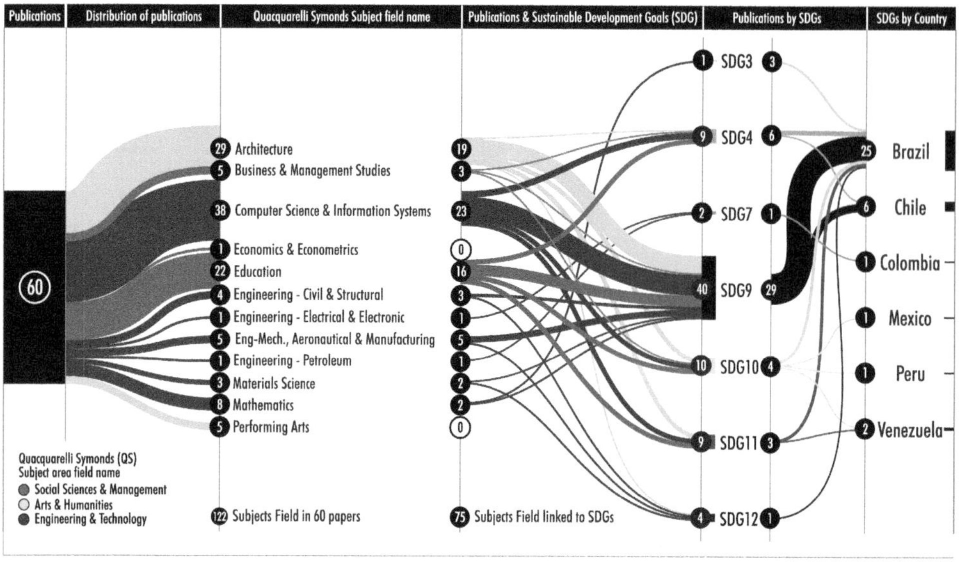

Figure 15.16 Scopus publications related to SDGs and Digital Fabrication.

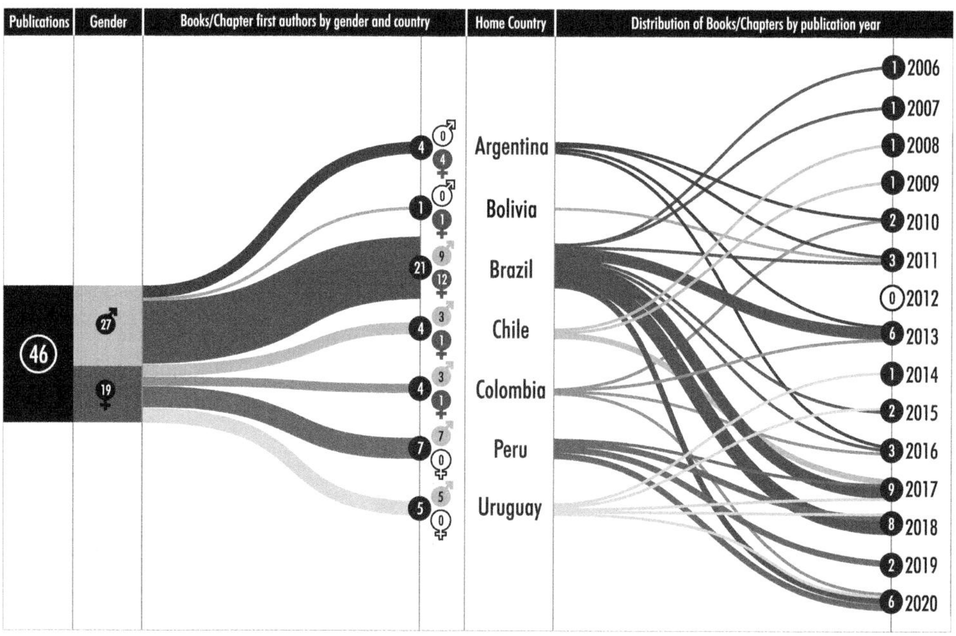

Figure 15.17 Books & Catalogs about Digital Fabrication. Distribution by country and year.

Mapping Making Culture in Digital Fabrication

The development of maker culture in Latin America was shaped largely by computational design workshops and hands-on learning. This set of diagrams focuses on formal and informal events linked to digital fabrication in the Global South. Workshops became the

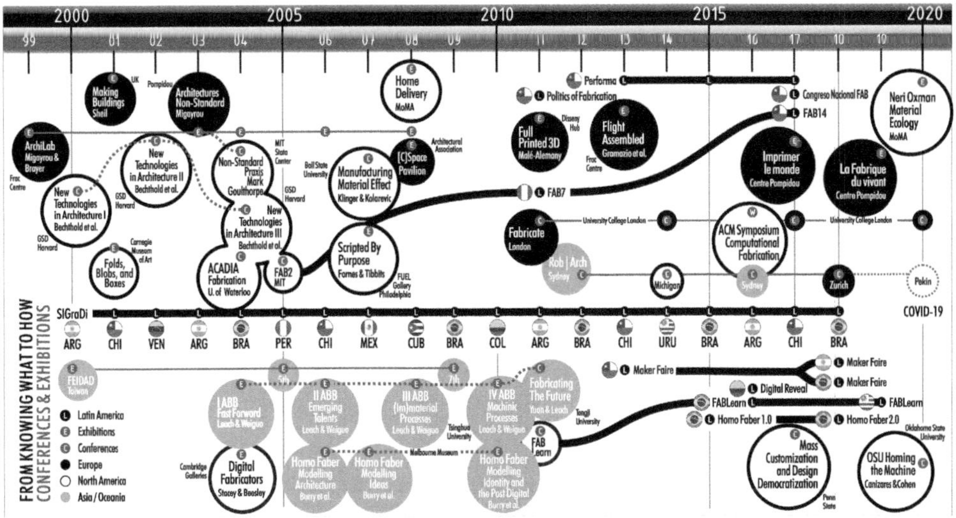

Figure 15.18 Digital Fabrication. Global mapping of events and exhibitions 2000–2020.

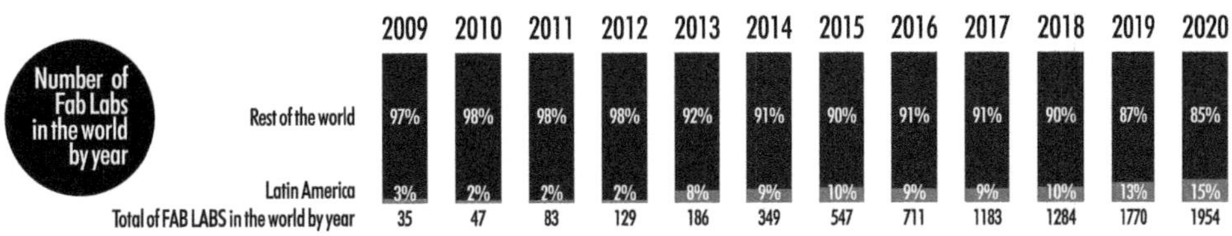

Figure 15.19 Number of Fab Labs by year in the world, and the percentage of Latin America participation.

key resources for blending knowing and making and, unlike in the Global North, programming in Latin America began with design workshops and later integrated digital fabrication (Figure 15.18).[21]

In 2020, the Fab Foundation worldwide network consisted of 1,954 Fab Labs. Two hundred and eighty six (15%) of these corresponded to Latin America (Figure 15.19), South America (12%), and Central/North America (3%). The first Latin American country to participate in this global network was Costa Rica in 2002 at the initiative of Bakhtiar Mikhak of the Grassroots Invention Group of the MIT Media Lab together with the CBA-MIT project of Neil Gershenfeld. This initiative linked Costa Rica, El Salvador, and Mexico with the MIT Digital Nations program, which supported the creation and promotion of local inventions with appropriate technology in developing countries (Rodriguez, 2003).[22] However, only the Costa Rica Institute of Technology remained in the project. The number of laboratories grew with the Fab Academy project in 2009 that reached South America and Mexico. However, Central America continued to have implementation limitations. In 2016, the architect Robert Garita from Fab Lab Veritas in Costa Rica and the Fab Lab Association of the same country created the "Build it up" program. This initiative sought to integrate Central America and the Caribbean. In 2017, the Fab

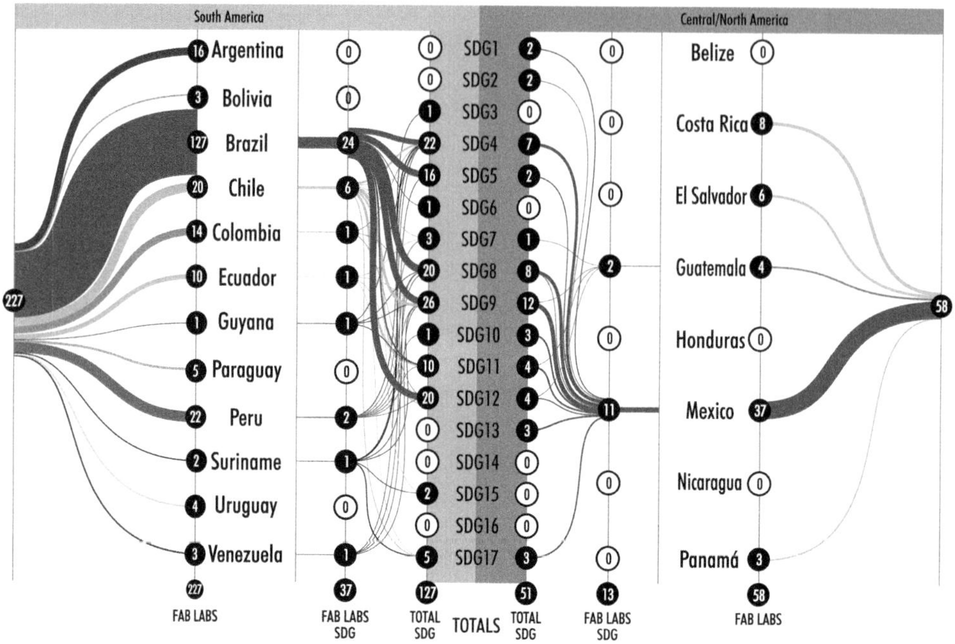

Figure 15.20 Number of Fab Labs registered at the Fab Foundation and UN SDGs (2000–2020).

Foundation chose this initiative as the Central American Node of the Fab Academy. Despite this initiative, three countries still do not have infrastructure associated with the Fab Foundation (Belize, Honduras and Nicaragua) (Figure 15.20). Additionally, many Fab Labs in the Global South incorporate the UN's SDGs as part of their missions. They have sustainability as a main focus. Figure 15.17 shows statistics from the Fab Foundation, featuring the 227 laboratories installed in South America until 2020 and the 37 laboratories that declared specific outcomes related to SDGs.

The exhibitions *Homo Faber 1.0 Digital Fabrication in Latin America, Homo Faber 2.0 Politics of Digital Fabrication*, and *Homo Faber 3.0 Appropriations from Latin America* have been some of the most important curatorial events for the new generation of Latin American designers and researchers focused on the digital fabrication from architecture. These exhibitions and their catalogs defined categories that expose the most relevant digital fabrication in Latin America, from the scale of the object to architecture. Figure 15.21 illustrates the origin and number of laboratories that participated in these series, with projects that become the starting point for inspiration and new visions.

In 2019, we began to analyze the activities of the Mobile Fab Labs and those that implemented robots in architecture.[23] Though new initiatives, these activities demonstrate the variety of formats being explored. Mobile Fab Labs appeared with a precedent in the Mobile Fab Lab of the Fab Foundation inaugurated in 2007. As the name suggests, they are aimed at communities that could benefit from temporary technological infrastructures. With Mobile Fab Labs, the technology and knowledge that was originally only accessible in universities can reach a wider audience.

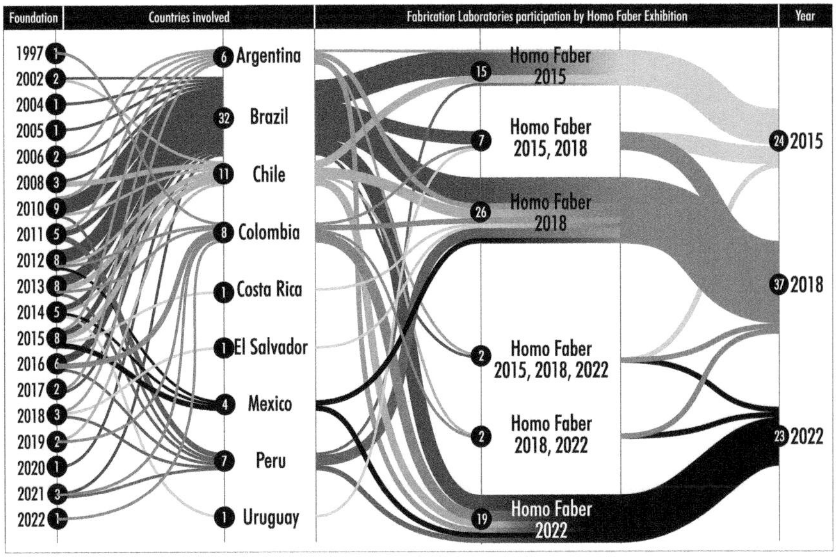

Figure 15.21 Fabrication Laboratories in Homo Faber Exhibitions (2015–2022) by country.

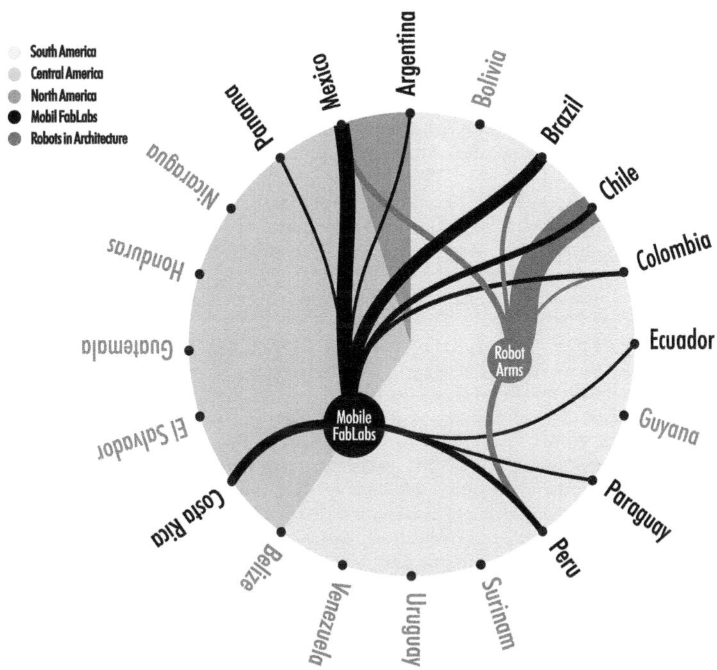

Figure 15.22 Mobile Fab Labs and Robots in Latin America by numbers.

Robotic implementations are also increasing. However, the scale of these instruments demands a lot of resources. The cost of implementing laboratories is a major hindrance in many countries. For example (Figure 15.22), countries like Mexico, with industrial manufacturing plants for auto parts, facilitated a straightforward incorporation of robotic arms at engineering schools. These industries donated equipment that had completed its industrial lifecycle to universities in that country. Elsewhere, schools of architecture must use their own resources or competitive funds to respond to academic initiatives related to automation. In our study, we identified five countries, Brazil, Chile, Colombia, Mexico, and Peru that have been implementing the use of robots in architecture schools. Mainly in universities with a track record in computational design and lines of research that created opportunities in academia and their community.

Acknowledgments

This case study originates from discussions and research with different academics with whom we share the same interests on emerging technology in Latin America. Without their initial contribution, this evolution in progress could not be revealed: Frederico Braida, Daniel Cardoso Llach, Gabriela Celani, Cristina Dreifuss-Serrano, Rodrigo García Alvarado, Rodrigo Scheeren, David Sperling, and Macarena Valenzuela-Zubiaur. This work has been financed by the Research Department of the Peruvian University of Applied Sciences with research project number C-13-2019.

Notes

1 Granger, Frank. Vitruvius on architecture. Books 1–5. (Original text and English translation). Cambridge: Harvard Loeb Classical Library, 1998 [1931]. 6.
2 Granger, 7.
3 Flusser, Vilém. Die Fabrik [The Fabric]. In Vom Stand der Dinge. Eine kleine Philosophie des Design Herausgegeben von Fabian Wurn. Frankfurt; Steidl Verlag, 1993 [1991]. 68–75.
4 Gershenfeld, Neil. *FAB. The Coming Revolution on Your Desktop-from Personal Computers to Personal Fabrication.* New York; Basic Books, 2005.
5 Picón, Antoine. *Digital Culture in Architecture: An Introduction for the Design Professions.* Basel: Birkhäuser, 2010.
6 Basalla, George. *The Evolution of Technology.* Cambridge: Cambridge University Press, 1988.
7 Cortada, James W. *The Digital Hand. How Computers Changed the Work of American Manufacturing, Transportation, and Retail Industries.* Oxford: Oxford University Press, 2004. 44.
8 Cardoso Ll., Daniel. *Builders of the Vision. Software and the Imagination of Design.* New York: Routledge, 2015.
9 Research by Sperling et al., (2019), positioned digital fabrication in Latin America in 2006, showing its frequency in the top five rank keywords of the SIGraDi conferences until 2020. In his PhD thesis, Sheeren (2021) analyzed and characterized 116 laboratories in the region, categorizing their production and leaders of their management. In addition to these significant works, the curatorship of Latin American explorations in digital fabrication has been shown in Asia, the United States, and Latin America itself. These exhibits and their catalogs evidenced experimentation scales and categories that exposed their development. With these categories and the visibility of others, new research has appeared on the influence of Fab Labs on gender and leadership; vulnerable groups (kids and women), artisans, and micro-entrepreneurs. Also in this context, an important work by the team of van der Hijden et al. (2017) started at Fab13 in Chile, which identifies Fab Labs connected to the United Nations Sustainability Development Goals (UN SDGs). There is also research that identifies the reduction of technological gaps when infrastructure moves to the periphery with Mobile Fab Labs; and the programming of trajectory and automation of tasks with robotic fabrication.
10 Gershenfeld, Neil. "Overview." NSF Site Visit CCR-0122419, Boston, USA, August 2, 2002. https://cba.mit.edu/events/02.09.NSF/index.html.
11 Fab Foundation. "Getting Started with Fab Labs". https://fabfoundation.org/getting-started/.
12 Montagú, Arturo, Kos, José, Rodriguez, Diana, Stipech, Alfredo, and Rodrigo, García. "Digital Design Curriculum. Developments in Latin America – A Field Report." *19th eCAADe Conference Proceedings.* 202–206. Helsinski, Finland, August 29–31, 2001. http://papers.cumincad.org/data/works/att/2d5c.content.pdf. 202.

13 Herrera, Pablo C., and Juárez, Benito. "Perspectivas en los laboratorios de Fabricación Digital en Latinoamérica." *16th Conference SIGraDi*, 285–289. Fortaleza, Brazil, 13–16, November, 2012. http://papers.cumincad.org/data/works/att/sigradi2012_395.content.pdf.

14 Stokes, Lowery. *New Territories. Laboratories for Design, Craft and Art in Latin America.* Madrid: Museum of Arts and Design and TURNER, 2014.

15 Gaetano, Paula. "Introduction. Unsettled Artifacts: Technological Speculations from Latin America. Art Gallery," *Leonardo* 50, no.4 (2017):410–413. https://doi.org/10.1162/LEON_a_01459.

16 Between 2002 and 2022, 1,186 projects linked to "Additive Manufacturing" received US$592M from the NSF. Of these, 33% (391 projects) will continue to execute this financing between 2023 and 2027. Of the total, 217 educational institutions led 86% (1.018 projects). On the other hand, between 2007 and 2022, NSF funded 25 projects connected to "Fab Labs" with US$18.1M and, between 2015 and 2021, NSF funded 30 projects linked to "Digital Fabrication" with US$17.3M.

17 García, Rodrigo, Henriques, Gonçalo, and Chiarella, Mauro. "Co-creating together: Ibero-American emergent collaborations in architecture and digital manufacture," *Arquitectura del Sur* 40, no.61 (2022): 118–133. https://doi.org/10.22320/07196466.2022.40.061.08.

18 Schork, Tim. "Option Explicit: Scripting as Design Media." *What's Matters? First International Conference on Critical Digital Conference Proceedings.* 41–46. Boston, MA: Harvard Graduate School of Design, 2008. 46.

19 Herrera, Pablo C., Caycho, Vaneza, and Valenzuela-Zubiaur, Macarena. "Women in the Fab Lab ecosystem (2008-2021). From Fab Academy to the Fab Lab Research Conferences". *Proceedings of the Fab 16 Research Papers Stream*, 271–280. Montréal, Canada, 9–12 August 2021c. https://doi.org/10.5281/zenodo.5169858.

20 See Cardoso, Ll., Daniel and Burbano, Andres. "Other Computations: Digital Technologies for Architecture from the Global South," *Dearq* 27 no.1 (2020): 6–19. https://doi.org/10.18389/dearq27.2020.01; Noel, Vernelle. "Computational Regionalism: De-Familiarization of Tectonics in the Wire-Bending Craft." *International Journal of Architectural Computing* 20, no. 2 (June 2022): 277–296. https://doi.org/10.1177/14780771221097685; and Voto, Cristina, Sperling, David, and Martin-Iglesias, Rodrigo. "Digitalidades del Sur". *AREA - Agenda De Reflexión En Arquitectura, Diseño Y Urbanismo* 29, no. 1 (2023): 1–4 https://publicacionescientificas.fadu.uba.ar/index.php/area/article/view/1922.

21 In 2002, Gabriela Celani completed her Ph.D. at MIT and implemented AutoLISP in Brazil. Celani (2002) taught experimental courses during the first semester of 2002 at the "Universidad de São Paulo" and at theUNICAMP. In these courses and workshops, programming allowed students to simulate designs and experiment with materiality. In 2007, UNICAMP implemented the LAPAC digital fabrication laboratory.

In 2006, Claudio Labarca and Verónica Arcos from the Pontificia Universidad Católica de Chile conducted the workshop "Generative Algorithms" using Maya Embedded Language (MEL) Script, which was attended by other Chilean scholars Arturo Lyon and Kristof Crolla who would graduate from the Architectural Association Design Research Laboratory (AA DRL) and work for Zaha Hadid Architects. In 2007, Labarca co-founded the "Taller de Ensable Digital" with Arturo Lyon and Diego Pinochet.

In November 2006, I organized with Pedro Soza from the Universidad de Chile, the Computational Design Solutions Workshops using Rhinoscript at the 10th SIGraDi conference. K. Griffith, J. Snavely, and D. Cardoso Llach were Ph.D. and Master's students at MIT and were responsible for the development of this implementation. A follow-up workshop included professors at the Universidad de Chile in 2007, Peru (2008) and Daniel Cardoso Llach with Skylar Tibbits in Colombia (2009). As a result, the Universidad Peruana de Ciencias Aplicadas in Peru and Universidad de los Andes in Colombia, which hosted these experiences, now have standards for their digital fabrication laboratories and specialized courses in visual programming for trajectory control of industrial robot arms.

In Brazil, Gabriela Celani and Regiane Pupo gave rapid prototyping workshops between 2007 and 2008. In 2008, Underléa Bruscato managed the "Architecture and Digital Fabrication Workshop" at the Universidade do Vale do Rio dos Sinos (Brazil). This workshop was integrated with the UBB (Chile), Universidad Católica de Córdova (Argentina), and the Pontificia Universidad Católica de Chile, to evaluate learning digital fabrication with new equipment and online learning.

In 2009, Marcela Godoy and the UTFSM Collective Bangs organized a 2009 RhinoScript Workshop with Marc Fornes. Students from this workshop Francisco Calvo and Katherine Cáceres (Chile) founded the Dum Dum Lab in 2010. Another student Francisco Quitral co-directs with Luis Felipe Gonzales different implementations of robotics in architecture. In 2008, Collective Bang published a group of essays on digital fabrication experiences, projects, and research by Rodrigo García, Luis Felipe Gonzáles, Sergio Araya, and Jorge Godoy. The authors and alumni influenced the Chilean history of digital fabrication.

In 2006, a small tool called Explicit History became the origin of Grasshopper plugin (2008), a group of visual programming components for Rhinoceros 4.0 created by David Rutten as a counterpart to Bentley's Generative Components. Grasshopper changed the way of approaching programming and consequently made it easier to diagram digital fabrication processes in a parametric way without using or learning lines of code. Between 2008 and 2009, the preference for visual programming was intense in Latin America over scripting (Herrera, 2011). SIGraDi and workshops mediated between visual programming, techniques, and digital fabrication infrastructure available in universities.

22 Rodríguez, Wendy. "RAI como proyecto tecnológico". Primer Taller Latinoamericano Ciencia, Comunicación y Sociedad. San José, Costa Rica, 24–26 November, 2003. https://www.cientec.or.cr/comunicacion/ponencias/WendyRodriguez.pdf.

23 Herrera, Pablo C., and Valenzuela-Zubiaur, Macarena. "The impact of the democratization of technologies in vulnerable groups, artisans, and micro-entrepreneurs. The role of Mobile Fab Labs". *2022 IEEE International Symposium on Technology and Society (ISTAS)*. Hong Kong, November 10–12, 2022b.

Bibliography

Alvarez, Natalia, Bernal, Marcelo, and Cáceres, Katherine. "Evolution and Projection of Computational Design Theories: Generation, Analysis, Selection and Fabrication." *24th Conference SIGraDi*, 120–127. Online Conference. November 18–20, 2020.

Basalla, George. *The Evolution of Technology*. Cambridge: Cambridge University Press, 1988.

Cardoso Ll., Daniel. *Builders of the Vision. Software and the Imagination of Design*. New York: Routledge, 2015.

Cardoso, Ll., Daniel and Burbano, Andres. "Other Computations: Digital Technologies for Architecture from the Global South," *Dearq* 27 no.1 (2020): 6–19. https://doi.org/10.18389/dearq27.2020.01.

Celani, Gabriela, and Pupo, Regiane. "Rapid Prototyping and Digital Fabrication for Architecture and Building: Definitions and State of the Art in Brazil," *Cadernos de Pós- Graduação em Arquitetura e Urbanismo* 8, no.1 (2008): 31–41.

Celani, Gabriela. "Shortcut to the Fourth Industrial Revolution: The case of Latin America," *International Journal of Architectural Computing* 18, no.4 (2020): 320–334. https://doi.org/10.1177/147807712094219.

Fab Foundation. "Getting Started with Fab Labs". https://fabfoundation.org/getting-started/.

Flusser, Vilém. Die Fabrik [The Fabric]. In Vom Stand der Dinge. Eine kleine Philosophie des Design Herausgegeben von Fabian Wurn. Frankfurt; Steidl Verlag, 1993 [1991], 68–75.

Gaetano, Paula. "Introduction. Unsettled Artifacts: Technological Speculations from Latin America. Art Gallery," Leonardo 50, no.4 (2017):410–413. https://doi.org/10.1162/LEON_a_01459.

García, Rodrigo, Henriques, Gonçalo, and Chiarella, Mauro. "Co-creating Together: Ibero-American Emergent Collaborations in Architecture and Digital Manufacture," *Arquitectura del Sur* 40, no.61 (2022): 118–133. https://doi.org/10.22320/07196466.2022.40.061.08.

Gershenfeld, Neil. FAB. *The Coming Revolution On Your Desktop-From Personal Computers to Personal Fabrication*. New York: Basic Books, 2005.

Gershenfeld, Neil, Gershenfeld, Alan, and Cutcher-Gershenfeld, Joel. *Designing Reality. How to Survive and Thrive in the Third Digital Revolution*. New York: Basic Books, 2017.

Herrera, Pablo C. "Computational Design Solutions in architectural education: The use of script language to design complex surfaces". *11th Conference SIGraDi*, 95–101. Mexico, 23–25, 2007.

Herrera, Pablo C. "Towards an identity: Digital Fabrication in Latin America." *AA Politics of Fabrication Laboratory. Symposium International Conference*, Valparaiso, Chile, May 13, 2011.

Herrera, Pablo C., and Juárez, Benito. "Perspectivas en los laboratorios de Fabricación Digital en Latinoamérica." *16th Conference SIGraDi*, 285–289. Fortaleza, Brazil, 13–16, November, 2012. http://papers.cumincad.org/data/works/att/sigradi2012_395.content.pdf.

Herrera, Pablo C. "Scripting and Digital Fabrication in Artisanal Processes: The Case of South America". *Cross Americas: Probing Disglobal Networks. ACSA International Conference*, 357–362. Santiago, Chile, June 29 – July 1, 2016 https://doi.org/10.35483/ACSA.Intl.2016.64.

Herrera, Pablo C., and Braida, Frederico. "Digital Technologies in Latin American Architecture - A Literature Review from the Third to the Fourth Industrial Revolution" *19th eCAADe Conference Proceedings*. 431–440. Porto, Portugal, September 11–13, 2019. https://doi.org/10.52842/conf.ecaade.2019.1.431.

Herrera, Pablo C., Caycho, Vaneza, and Valenzuela-Zubiaur, Macarena. "Women in the Fab Lab ecosystem (2008-2021). From Fab Academy to the Fab Lab Research Conferences". *Proceedings of the Fab 16 Research Papers Stream*, 271–280. Montréal, Canada, 9–12 August 2021. https://doi.org/10.5281/zenodo.5169858.

Herrera, Pablo C., Sperling, David, Henriques, Gonçalo, and Vizioli, Simone. *SIGraDi #25 Sociedad Iberoamericana de Gráfica Digital 1997-2021*. São Carlos: Instituto de Arquitetura e Urbanismo, Universidade de São Paulo. https://doi.org/10.11606/9786586810288.

Herrera, Pablo C., Dreifuss-Serrano, Cristina, Valenzuela-Zubiaur, Macarena, and Caycho, Vaneza. "Fab Lab-based learning: An environment to promote Women and Leadership in Engineering Education". *2022 IEEE Global Engineering Education Conference (EDUCON)*, 1336–1345. Tunisia, March 28–21, 2022. https://doi.org/10.1109/EDUCON52537.2022.9766775.

Herrera, Pablo C., and Valenzuela-Zubiaur, Macarena. "The impact of the democratization of technologies in vulnerable groups, artisans, and micro-entrepreneurs. The role of Mobile Fab Labs". *2022 IEEE International Symposium on Technology and Society (ISTAS)*, Hong Kong, November 10–12, 2022.

Herrera, Pablo C., Scheeren, Rodrigo and Sperling, David. *Homo Faber 3.0. Appropriation of Digital Fabrication in Latin America*. Lima: Editorial Universidad Peruana de Ciencias Aplicadas, 2023.

Iriarte, Pablo. "Exhibition Projects. Digital Craft in Semi-peripheral Nations." In *Projects Catalog of the 38th Annual Conference of the Association for Computer Aided Design in Architecture*, edited by Phillip Anzalone, Marcella del Signore, and Andrew John Wit, 52–73. México: Universidad Iberoamericana, 2018.
Leach, Neil, and Weiguo, Xu. *(Im)material Processes: New Digital Techniques for Architecture. Architecture Biennial Beijing*. Beijing: Tsinghua University, 2008.
Leach, Neil, and Weiguo, Xu. *Machinic Processes. Architecture Biennial Beijing*. Beijing: Tsinghua University, 2010.
Leach, Neil and Yuan, Philip. *Fabricating the Future*. Tongji: Tongji University Press, 2012.
Martens, Bob, and Turk, Ziga. "Working experiences with a Cumulative Index on CAD: CUMINCAD". 327–333. eCAADe 1999. Liverpool, UK, 15–17 September, 1999.
Mikhak, Bakhtiar, Lyon, Christopher, Gorton, Tim, Gershenfeld, Neil, McEnnis, Caroline, and Taylor, Jason. "Fab Lab: An Alternate Model of ICT for Development". Cambridge: Grassroots Invention Group MIT Media Laboratory, 2002. https://cba.mit.edu/events/03.05.fablab/fablab-dyd02.pdf.
Miyazaka, Elza, and Fabrício, Márcio. "Digital fabrication in Brazil. Academic production in the last decade". *16th Conference CAAD Futures 2015*. 421–433. São Paulo, Brazil, July 8–10, 2015.
Montagú, Arturo, Kos, José, Rodriguez, Diana, Stipech, Alfredo, and Rodrigo, García. "Digital Design Curriculum. Developments in Latin America – A Field Report." *19th eCAADe Conference Proceedings*. 202–206. Helsinski, Finland, August 29–31, 2001. http://papers.cumincad.org/data/works/att/2d5c.content.pdf.
Noel, Vernelle. "Computational Regionalism: De-Familiarization of Tectonics in the Wire- Bending Craft," *International Journal of Architectural Computing* 20, no. 2 (June 2022): 277–296. https://doi.org/10.1177/14780771221097685.
Peña, Peña, Lal, Bhavya, and Micali, Maxwell. "U.S. Federal Investment in the Origin and Evolution of Additive Manufacturing: A Case Study of the National Science Foundation," *3D Printing and Additive Manufacturing* 1, no.4 (2014): 185–193. http://doi.org/10.1089/3dp.2014.0019.
Picón, Antoine. *Digital Culture in Architecture: An Introduction for the Design Professions*. Basel: Birkhäuser, 2010.
Rodríguez, Wendy. "RAI como proyecto tecnológico". Primer Taller Latinoamericano Ciencia, Comunicación y Sociedad. San José, Costa Rica, 24–26 November, 2003. https://www.cientec.or.cr/comunicacion/ponencias/WendyRodriguez.pdf.
Scheeren, Rodrigo, Herrera, Pablo C., and Sperling, David. *Homo Faber 2.0: Politics of Digital in Latin America*. São Carlos: Instituto de Arquitetura e Urbanismo USP, 2018.
Scheeren, Rodrigo. "Fabricação digital na América do Sul: laboratórios, estratégias, processos e artefatos para o design, a arquitetura e a construção," PhD diss., (Universidade de São Paulo, 2021). https://doi.org/10.11606/T.102.2021.tde-05042022-173034.
Scheeren, Rodrigo. "Outros olhares para a práxis no design e na arquitectura. Notas sobre o principio do uso das tecnologías de fabricação digital na América do Sul." AREA - Agenda De Reflexión En Arquitectura, Diseño Y Urbanismo 29, no. 1 (2023): 1–14 https://publicacionescientificas.fadu.uba.ar/index.php/area/article/view/1918.
Sennett, Richard. *The Craftman*. New Haven, CT: Yale University Press, 2008.
Sperling, David, Herrera, Pablo C., and Scheeren, Rodrigo. "Migratory Movements of Homo Faber: Mapping Fab Labs in Latin America." In *CAAD Futures 2015*, edited by Gabriela Celani, David Sperling, and Juárez Franco, 405–421. Berlin: Springer, 2015a. https://doi.org/10.1007/978-3-662-47386-3_22.
Sperling, David, Herrera, Pablo C., and Scheeren, Rodrigo. Homo Faber. *Digital Fabrication in Latin America*. São Carlos: Instituto de Arquitetura e Urbanismo, USP, 2015b.
Sperling, David, Vizioli, Simone, Botasso, Gabriel, Tiberti, Mateus, Santana, Eduardo, and Sígolo, de Oliveira Ordonho. "Crossing Timelines - Main research topics in the histories of eCAADe and SIGraDi" *19th eCAADe Conference Proceedings*. 407–416. Porto, Portugal, September 11–13, 2019. https://doi.org/10.52842/conf.ecaade.2019.1.407.
Sperling, David, Herrera, Pablo C., and Scheeren, Rodrigo. "Fabricating (Other) Computations: Digital Fabrication and Technological Appropriation in Latin America," *Dearq*, no.27 (2020): 76–87. https://doi.org/10.18389/dearq27.2020.06.
Stokes, Lowery. *New Territories. Laboratories for Design, Craft and Art in Latin America*. Madrid: Museum of Arts and Design and TURNER, 2014.
van del Hijden, Pieter, Bassi, Enrico, and Caycho, Vaneza. "How to align your Fab Lab / Makerspace with the U.N. Sustainable Development Goals (SDGs)". https://gitlab.fabcloud.org/fl-management/fl-sdgs/sdg-workshop.
Voto, Cristina, Sperling, David, and Martin-Iglesias, Rodrigo. "Digitalidades del Sur". AREA - Agenda De Reflexión En Arquitectura, Diseño Y Urbanismo 29, no. 1 (2023): 1–4. https://publicacionescientificas.fadu.uba.ar/index.php/area/article/view/1922.
Walter-Hermann, Julia, and Büching, Corinne. *FabLab. Of Machines, Makers, and Inventors*. Bielefeld: Transcript, 2013.
Yuan, Philip, Menges, Achim, and Leach, Neil. *Digital Fabrication*. Shanghai: Tongji University Press, 2017.

A Carrier Bag of Tools for Computational Feminism

Shelby Doyle

Introduction

In the "The Carrier Bag Theory of Fiction," author Ursula Le Guin retells the story of human origins by describing the first technology as a vessel rather than a spear.[1] Le Guin's "carrier bag theory" draws upon the work of anthropologist Elizabeth Fisher who posits that the "earliest cultural invention must have been a container" such as a bag to hold gathered food or a sling to carry children, rather than a weapon of hunting or war—a common assumption of evolutionary and technologic narratives.[2] Le Guin wrote that the "carrier bag theory" created space for humanity and everyday innovations in her science fiction rather than only space for the hard sciences.[3] In a more recent piece, she goes on to suggest that technology is socially constructed and is "the active human interface with the material world," meaning that "we can all learn. That's the neat thing about technologies. (Technology) is what we can learn to do…"[4] As an architect, educator, and researcher, I find this definition of technology useful because it is expansive in its inclusion of human ingenuity across time and beyond machines.

If architects engaged in digital fabrication practices define technology in terms of human culture rather than scientific objectivity, as Le Guin suggests we do, we can begin to advance toward creating a "technocultural" framework for our work.[5] I use the term "technocultural" narrowly to describe the origin stories, digital fabricators, exchange, and often rely upon, about humans co-creating with machines. The "carrier bag theory" is a reminder that such stories are value propositions. Architecture and its technologies

 DOI: 10.4324/9781003296522-19

can be enriched if we acknowledge that our technologic origins are both/and: both a vessel, and its associated act of caretaking, and a spear, and its associated acts of violence.

In the context of this book, I take on "home" and "homing the machine" as a search for alternative, technocultural origin stories. More specifically, I approach "homing" as an opportunity to contextualize digital fabrication within alternative creation myths and lineages that expand how we, architects, think about—and with—technology, as well as who participates in technological practices. In its present "home," digital fabrication in architecture excludes many people, ideas, and forms of knowledge. By acknowledging multiple technocultural origins, we can create a field that provides a "home" to a greater diversity of ideas and people.

The origins of the Iowa State University (ISU) Computation and Construction Lab (CCL) are sympathetic with Le Guin's carrier bag theory. The CCL is, like many digital fabrication labs, a physical space that houses machines and robotics. However, the Lab is also a home to pedagogical experiments, fabricated objects, and theoretical ambitions organized around the ideas of "computational feminism." Computational feminism is a tactic that operates relentlessly upon the cultures of architecture and computation using disciplinary tools: writing, teaching, coding, drawing, making, and building. In the following case study, I share projects from each of these categories to reflect on how architects and designers make with, attribute, and historicize computational knowledge and labor. My proposition is ultimately that "computational feminism" is one of the many technocultural origin stories needed to truly "home the machine."

Founding: Origins of the CCL

The origins of the CCL are political, personal, and practical. In 2015, I was hired under the ISU President's "High Impact Hire Initiative" to develop "innovative" work in digital fabrication and design-build at the ISU College of Design and Department of Architecture. The High Impact Hire Initiative "supported targeted faculty hiring in areas of strategic importance" to ISU; one such area was that of "innovation." ISU sought to develop cutting-edge teaching and research that would attract students in an increasingly competitive educational market.[6,7]

I approached my new position from various standpoints. As a former student of Neil Gershenfeld's *How to Make (Almost) Anything* course at the MIT Media Lab, I was working in the lineage of the Center for Bits and Atoms' "Fab Labs" and the emancipatory and democratizing potential of personal computing as a tool of creativity, imagination, and optimism.[8] As an architect, I was grappling with the seemingly immovable gender disparity in the upper echelons of the profession and the personal challenges I encountered as a woman operating in spaces of technology and construction.[9] As new tenure-track faculty, I was simultaneously designing new courses, onboarding fabrication machines, and establishing a research agenda about feminism and equity in architecture. These perspectives eventually coalesced under the rubric of computational feminism.

The ISU Computation and Construction Lab is housed in the former sound stages of Iowa Public Radio and Iowa Public Television on the ISU campus in Ames, Iowa. The space is a surprisingly good fit for a digital fabrication lab. The combined stages are 5,000 square feet of slab on grade floors, with high ceilings, no interior columns, and a high electric load—a legacy of the power needs of stage lights (Figure 16.1).

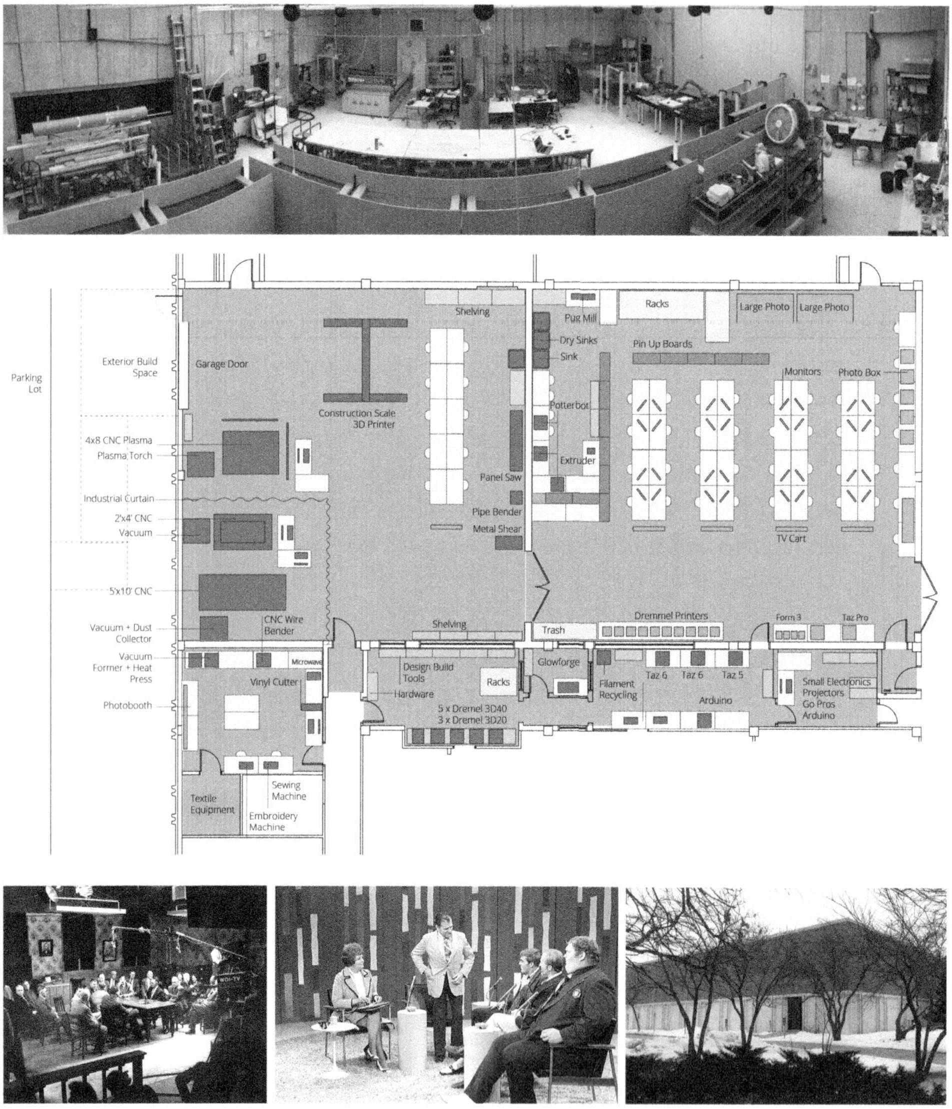

Figure 16.1 Top: Panorama of the Computation & Construction Lab in 2019. Middle: Floor plan of the CCL in 2022 indicating fabrication equipment and studio furniture. Bottom: Historic images of Iowa Public Television (Source: ISU Special Collections). Exterior of Communications Building.

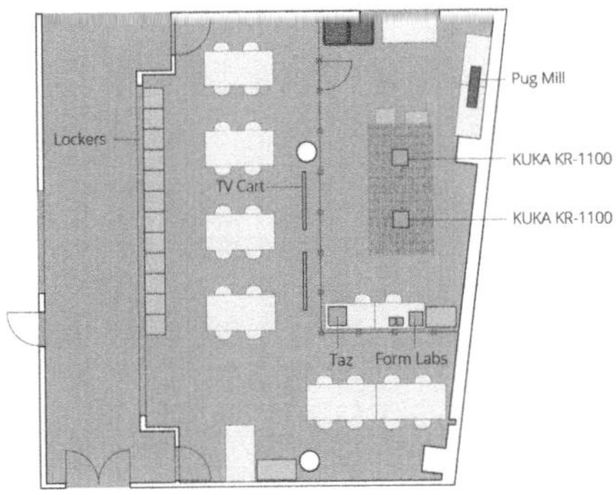

Figure 16.2 Left: Floor plan of the ARL in 2022 indicating fabrication equipment and studio furniture. Right: Image of author and Dai Le developing clay stamping workflows for the KUKA KR-1100 robotic arms. Photo by Christopher Gannon for ISU.

Today, the CCL houses $1 million of fabrication equipment and architectural technologies: Computer Numerical Control (CNC) routers, 3D printers, robotics, and construction tools, many of which now reside within the former control booths located behind the black glass to the left in the panorama in Figure 16.1. A few buildings away from the CCL is the Architectural Robotics Lab (ARL), a 1,500-square-foot outpost located in the ISU Student Innovation Center, designed by the American architecture firm, KieranTimberlake, and completed in 2021 (Figure 16.2).

Constructing both labs, the CCL and ARL, involved a substantial investment of space, time, and capital from ISU. Such an investment was only possible through administrative buy-in, as well as the time, and labor, of dozens of students and staff. As a "University of Science and Technology," ISU extols the technocratic virtues of innovation and entrepreneurship in its strategic plan.[10] Within this context, our adoption of the term "lab" for a space of architectural research is intentional yet ambivalent. Labs are deeply entangled in ongoing attempts across higher education in the United States to entrepreneurial-ize the humanities and scholarship in general.[11] Several scholars also note that the scientific laboratory inherits legacies of the Enlightenment and the European project of modernization and colonization, which in turn rely upon philosophies of rationality and truth. This myth of knowledge production through "objectivity" is at odds with Le Guin's conception of the "carrier bag." In her essay, "Situated Knowledges," the cultural theorist Donna Haraway terms such an epistemological myth "the god trick" or "seeing everything from nowhere."[12] Haraway's work generally highlights the ways in which scientific knowledge is often treated as a form of "universal truth" rather than conceived, constructed, and understood through social practices. She goes on to say that "we're living in a world of connections—and it matters which ones get made and unmade."[13] In other words, the things we make, how we make them, and how we use them to reveal the social structures and values of our culture, and, in this case, the discipline of digital fabrication in architecture.

Figure 16.3 SEKI is a collection of 80 custom ceramic vases 3D printed and glazed by Shelby Doyle, Erin Hunt, Ingrid Lilligren, and Kelly Devitt for the installation of Wendy Wintersteen, ISU's first female president.

My ISU colleagues and I adopted the term "lab" for both visibility and legitimacy. Our lab, after all, would participate within the context of a university that proudly announces that 57% of its graduates receive Science, Technology, Engineering, Math (STEM) degrees and recently decided to eliminate several graduate history programs.[14,15] Within this academic environment, there are clear tensions between the values of science and technology and the humanities as knowledge systems. The CCL attempts to hold these forms of knowledge together: to construct a pedagogical and research space focused on craft, computation, construction, and feminism under the guise of the "lab" at an institution of science and technology. The term "lab" is used both earnestly and as a decoy, smuggling Bauhausian studio traditions of clay, textiles, metal, stone, glass, and color (minus the exclusions of women from the architecture guild) into an institution that prioritizes the hard sciences.[16,17]

There are four continually evolving aims that inform the development of the CCL:

1 Testing ideas at full scale through built projects
2 Materializing computational practices through digital fabrication
3 Questioning education's role in replicating existing technological inequities
4 Pursuing technology in architecture as a space of gender equity rather than one of gender exclusion (Figures 16.3–16.6)

Computational Feminism as Carrier Bag

To choose computation and construction in architecture as a site for feminist discourse and action may seem like a strange decision. These are, after all, fields that have historically poor acknowledgment, integration, and representation of women.[18] To instrumentalize these inequities, the CCL draws from the concept of social determinism, that is, that society—and in this case the discipline of architecture—constructs the how, why, and who of technology. The counterargument, technological determinism, argues that

Figure 16.4 Left: Robotic painting during the Cyborg Sessions Workshop co-taught at ISU by Shelby Doyle, Erin Hunt, Nick Senske, and Leslie Forehand and culminating in an exhibition of the work and a public lecture by Madeline Gannon. Right: Disrupt/Displace an exhibition and performance for the 2016 Venice Biennale Sessions presenting North Dakota Access Pipeline a "Report from the Front." Fifty students constructed a grid in Iowa, packed it into 30 carry-on suitcases, and reconstructed it in Venice: co-taught by Shelby Doyle and Leslie Forehand. Photo by Alicia Pierce.

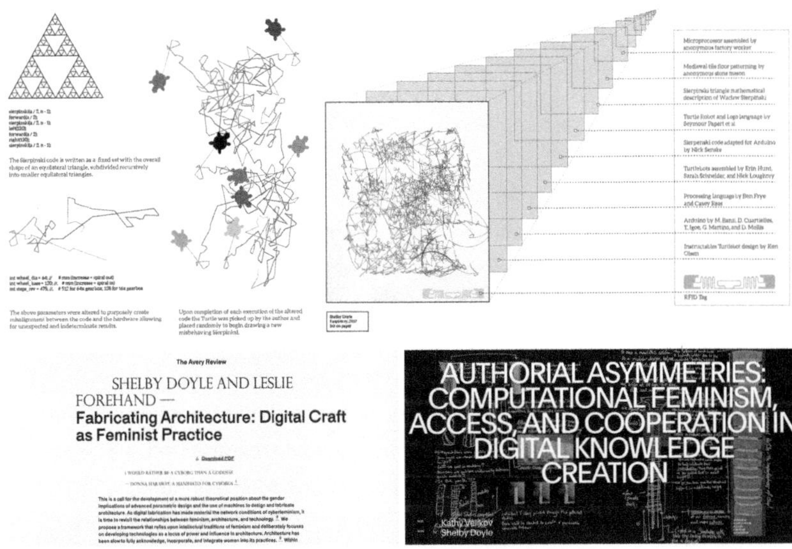

Figure 16.5 Top: Diagram from an essay in the International Journal of Architectural Computing co-authored with Nick Senske. The text questions how technology creates unique opportunities for attribution: to create space for people, AI's, and machines to be fully acknowledged for their intellectual and physical contributions to these objects and to be represented accurately in future histories of technology. Bottom Left: An essay by Avery Review published by Columbia University co-authored with Leslie Forehand which questions why digital fabrication and its recent histories—rely heavily upon narratives of biology and science—at the expense of craft traditions. Bottom Right: The arc of this work is discussed in an interview of Shelby Doyle by Kathy Velikov in Gradient Journal—Authorial Asymmetries: Computational Feminism, Access, and Cooperation in Digital Knowledge Creation.

Figure 16.6 Left: Plush, fabric cast concrete installation designed and fabricated by 12 students during a Spring 2021 undergraduate interdisciplinary studio, installed on campus during the COVID-19 pandemic. Right: IM_RU, 3D printed joints, wire and LEDs designed and fabricated by 16 students during Spring 2017 undergraduate interdisciplinary studio, installed at the Iowa Flyover Fashion Festival in Iowa City and then remade for the Iowa State Fair to represent ISU.

features of technology determine its use and it is the role of society (or architecture) to adapt to them.

To find a way into the feminist implications of computational design and construction, I conducted a crash course in feminist scholarship (a topic not directly addressed during the seven years of my architectural education). One conviction I found described by many feminist thinkers is that there is no hierarchy of oppressions. Rather, feminist scholars often invite the recognition that all oppressions—of gender, race, class, sexuality, religion, ability, and more—are connected; a term commonly referred to as "intersectionality."[19] For example, the author and social theorist Roxane Gay reminds us that an individual can simultaneously be privileged *and* marginalized, and what we need rather than perfect feminism is more effective ways to talk about unacknowledged power, untold stories, and unrecognized forms of knowledge.[20] I continue to use the term feminism, not as an act of denying a gender gradient, but rather to connect the lineages and legacies of feminist theory to present day technological discourses.

Connecting feminism to computational design and digital fabrication discourse is challenging because computational design is a slippery term and computational work is often presented as objective and therefore "extra-cultural." The MIT Design and Computation Group defines "computational design" as the "varied nature and practice of computation in architectural design, and the ways in which design meaning, intentions, and knowledge are constructed through computational thinking, representing, sensing, and making."[21] Computational design is a field that can be capacious, collective, inclusive, and radical—hallmark characteristics of feminist theory. However, though feminist theory has paid ample attention to the role of technology as a cultural practice, it appears that computational designers have demonstrated little interest in feminist thinking. For instance, in 2019, a search for "feminism" in the Cumulative Index of Computer Aided Architectural Design publications (CumInCAD) database and archive—an international digital repository of approximately 14,000 paper abstracts bringing together scholarship on architectural technology from major conferences and journals—returns only one search result. Perhaps it is this lack of intellectual overlap that perpetuates gender inequities—and not just gender representation—in computational design and digital fabrication.[22] To overcome this gap, I propose that architects and designers recognize feminism as an academic field with practices, methods, and theories from which architecture, computational design, and digital fabrication can learn (Figure 16.7).

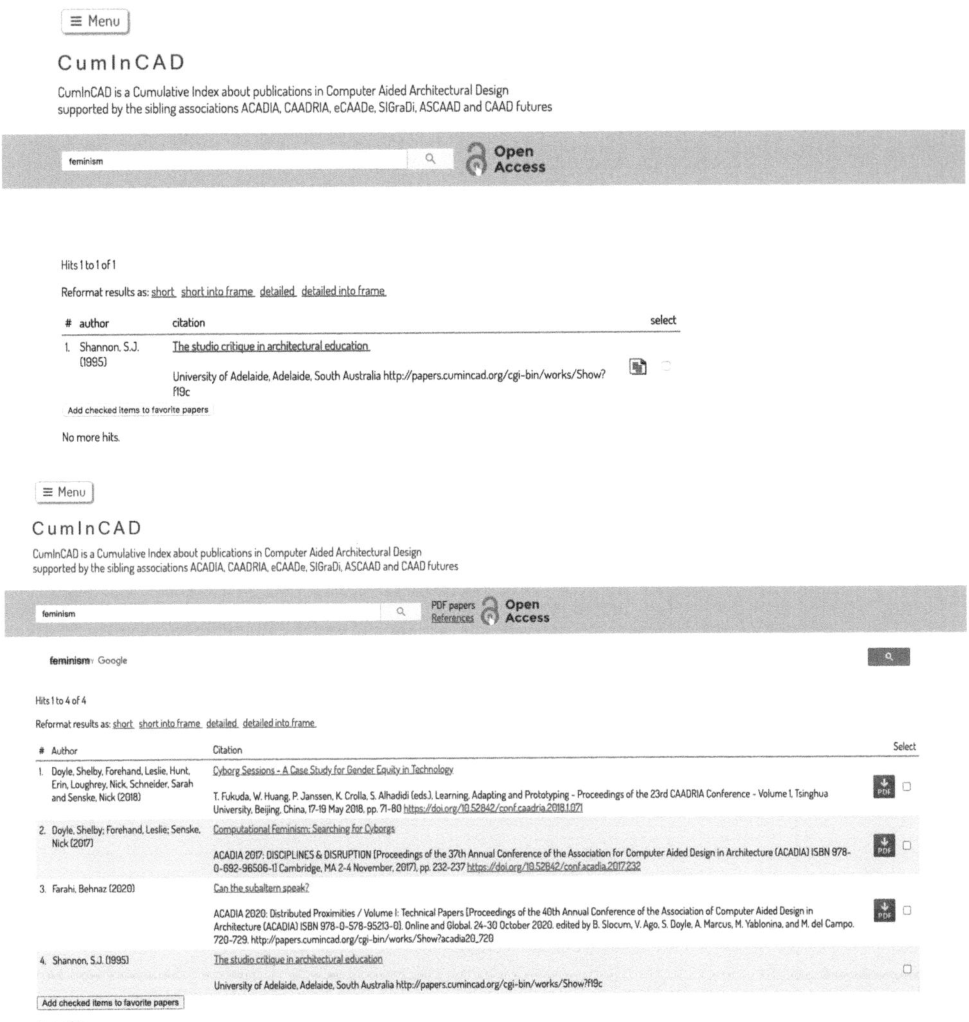

Figure 16.7 Screenshot of search results from the CumInCAD database in April 2017 and August 2022.

Computational feminism grew out of this gap-turned-intellectual space.[23] At the ACADIA 2017 Conference, Discipline and Disruptions, hosted by MIT, my colleagues, Nick Senske and Leslie Forehand, and I introduced our nascent theory:[24]

> Computational Feminism is a transdisciplinary field which grew out of the first twenty-five years of the digital turn. It continues to develop new theories on how politics of gender and other identity markers are interconnected to resulting processes of technical change, and the power relations of the globalized, material world. It is a descendant of the 1990s discourses of *technofeminism* and *cyberfeminism* that emerged in relationship to the development of network conditions and theories in architecture and urbanism.[25]

Computational feminism is a step toward integrating more ambitious aspects of feminist theory into computational design and digital fabrication discourse; in some ways, it is also the next step in work initiated at ISU. In the 1990s, ISU was the institutional home of architect Jennifer Bloomer. Bloomer approached what she viewed as the insufficiency of logical and rational structures in architecture with a radical feminist and interdisciplinary approach. For instance, she rejected the control, precision, and sterility of architectural drawings and called for them to be "irrational" or "subversive" instead. Bloomer's "dirty" drawings found parallels with the mess of childbirth and included bodily fluids, such as breast milk and menstrual blood.[26] These words—e.g., dirty or bloody—are not often (or ever?) found in computational design texts, likely cast out as terms that are too uncouth for a discipline fixated on solutions, efficiency, and control. But if we take on Bloomer's challenge, I think we might see that the irrational, the messy, and the subversive are valuable forms of knowledge and literacy that can counteract the bleaker aspects of computational thinking. Cultural theorist James Bridle argues for such subversion of our current computational world in their treatise the *New Dark Age: Technology and the End of the Future*:

> Computational thinking is predominant in the world today, driving the worst trends in our societies and interactions, and must be opposed by a real systemic literacy. If philosophy is that fraction of human thought dealing with what cannot be explained by the sciences then systemic literacy is the thinking that deals with a world that is not computable, while acknowledging that this is irrevocably shaped and informed by computation…[27]

At the CCL, computational design pedagogy follows Le Guin's proposition that technology can be learned, and really only exists, through the collective interplay of sharing and teaching. Therefore, technology is taught as a form of systemic design literacy that attempts to simultaneously teach what can be computed and reach beyond the limits of computation. In other words, at the CCL, we encourage students to resist technical, historical, and cultural biases using feminist tools of collective making, knowledge sharing, co-creating, and storytelling. Such tools are necessary to create new carrier bags and, ultimately, to enact a new digital turn.

Digital Turn: Sympoetic Tools for the Carrier Bag

The various digital turns that have occurred over the last few decades have made space for rethinking and resisting how architecture has been attributed, theorized, historicized, and practiced. The term, "digital turn," was first introduced by the architectural theorist Mario Carpo in his 1992 work, "The digital turn in architecture," in which he situates computational design in the long history of technology. In his more recent book, *The Second Digital Turn*, Carpo suggests that another turn in architecture has taken place. Carpo states that "the first digital turn in architecture changed our ways of making; the second changes our ways of thinking."[28,29]

As I mentioned above, the work at the CCL aspires to bring about a third digital turn in architecture, one that reorients the discipline's creation myths toward multiple origins and thrives in simultaneity, intersectionality, and collectivity. This new turn will be a sociocultural endeavor that builds on the tenets of computational feminism to organize technology through social exchange. Computational feminism reveals that feminist strategies have always existed in architecture—they do not need to be imagined into existence—therefore the fields of computational design and digital fabrication should be able to make immediate use of their principles. We can both learn from these histories and acknowledge the stories left untold. What, I think, remains is the CCL's mandate to operate relentlessly upon the cultures of architecture and computation using disciplinary tools: writing, teaching, coding, drawing, making, and building. These disciplinary tools enact the ambitions of computational feminism by broadening the methods and references that typically define architectural education and practice. With this expanded frame, computational feminism aims to collide with an architectural discipline that is sometimes hostile to both the existence of feminist ideas and the people those ideas represent.

Orienting feminism toward tools and tactics—code and machines, brooms and push carts, power drills and butcher block tables—means engaging with objects and things. Such an act of reorienting does not aim to abandon human and bodily concerns but rather to take seriously the labor of computational work and its entanglements with matter.[30] These are not extra-disciplinary anxieties; architecture, computational design, and digital fabrication participate in this world of tools, which, according to artist and theorist Katherine Behar writes, is "precisely one of exploitation...the world to which women, people of color, and the poor have been assigned under patriarchy, colonialism, and capitalism throughout history."[31] However, through the mechanisms of computational feminism, it might be possible that our disciplinary tools can be repurposed in service of more inclusive practices, tools, and ways of making.

To co-create with machines is to explicitly acknowledge the existence of collective intelligence and the impossibility of individualism. When we digitally fabricate, we are inextricably linked to the collective labor of those that have designed code, developed software, and built hardware.[32] As Donna Haraway notes, "nothing makes itself; nothing is really autopoietic or self-organizing." *Sympoesis*, or making-with, is a powerful alternative to *autopoesis*, or self-making; the former requires admitting that we are entangled in lineages, stories, cultures, and with each other.[33] She argues that "bounded ndividualism in its many flavors in science, politics, and philosophy has finally become *unavailable to think with*" Sympoesis is a method for navigating what Haraway calls the Chthulucene, our current epoch where human and nonhuman (machines, animals, plants, etc.) are entwined. This idea, that it matters what ideas we use to think of other ideas, is simple but important. I believe that *sympoesis* in computational design and digital fabrication can be a method to acknowledge computation's collective labor: from caretaking to coding.[34,35]

Conclusion

Computation is an idea that we use to think of other ideas; as such, it is an exceptionally powerful technology capable of both magnifying and reducing vast inequalities. As Bridle writes in the *New Dark Age*, "Technology is not mere tool making and tool use: it is the making of metaphors." They continue: "In making a tool, we instantiate a certain understanding of the world that, thus reified, is capable of achieving certain effects in that world."[36] It matters what stories we tell about technology and its creation myths as these lineages inform how we think about, and with, technology. For example, in *Race After Technology* sociologist Ruha Benjamin writes that "genealogies reflect and reproduce power relations. Citational practices are political, especially when talking about 'innovation' oppressed people and places are rarely cited for their many inventions."[37] As we know, history is not a simple meritocracy; it is a narrative of the past, written, and revised, or not written at all. Computational design, digital fabrication, and their recent histories rely heavily upon narratives of ecology, biology, and other hard sciences at the expense of the humanities, domestic traditions, and caretaking labor practices. My work and the work of the CCL attempt to shift these narratives.

The architectural histories of the future will be written from the records we keep today. These histories will represent how we collectively choose to value and attribute architectural work. I co-authored two articles with Nick Senske that examine how computation and digital fabrication can provide a record of its creation that opens up uniquely digital ways to track, cite, and source people and materials.[38,39] These essays reflect on the possibilities of and concerns over our ability to fully document digital labor, from text messages to mouse clicks. Computational labor is not only about the tools we use but also *how* we use them. In architecture, the interrogation of the world of matter and culture through computational methods is a cultural practice. But this cultural practice is not necessarily benign: computation can be largely invisible, an infrastructure only revealed by the culture it produces. Bridle writes

> The danger of this emphasis on the coproduction of physical and cultural space by computation is that it in turn occludes the vast inequalities of power that it both relies upon and reproduces. Computation does not merely augment, frame, and shape culture; by operating beneath our everyday, casually aware of it, it actually becomes culture.[40]

At the CCL by working with non-profits and small towns in Iowa, we work to imagine and build a computational world that welcomes discourse, dissent, and resistance to the inequalities computation can so easily, and often insidiously, reproduce.

Le Guin's "carrier bag theory" suggests that origin stories are value propositions about forms of knowledge and who gets to make with technology. The stories of the ISU CCL, "computational feminism," and the "sympoetic digital turn(s)" are some of the many origin stories needed to build new values for digital fabrication in architecture. In building these stories, the CCL strives to reveal how the politics of institutions can provide both opportunities and challenges for expanding disciplinary knowledge. Computational feminism asks that the world of digital fabrication, and architecture at large, grapple with its continued exclusions of many forms of knowledge and types of people. The "sympoetic digital turn(s)" requires conceding that nothing makes itself in architecture or otherwise. To turn and reorient digital fabrication and computational design toward multitudes of carrier bags means to actively construct a more inclusive and intersectional way of thinking-with computation in architecture; in other words, to truly "home the machine."[41]

Notes

1 Ursula Le Guin, *The Carrier Bag Theory of Fiction* (Barcelona: Terra Ignota, 1986).
2 Elizabeth Fisher, "The Carrier Bag Theory of Evolution" in *Worm's Creation: Sexual Evolution and the Shaping of Society* (New York: McGraw-Hill, 1980).
3 Ursula Le Guin, *The Carrier Bag Theory of Fiction* (Barcelona: Terra Ignota, 1986).
4 Ursula K. Le Guin, "A Rant About 'Technology,'" 2004. http://ursualkleguinarchive.com/Note-Technology.html.
5 The term technoculture is a neologism popularized by Penley, Constance, and Andrew Ross, eds., *Technoculture*. Vol. 3 (Minneapolis: U of Minnesota Press, 1991).
6 Stevens, Rob. "Presidential initiative sets sights on high-impact faculty hires", 2013. https://www.inside.iastate.edu/article/2013/09/12/hire. Accessed 5 December 2022.
7 Boeckenstedt, J. "Will your college survive the demographic cliff." *Chronicle of Higher Education* (2022). https://www.chronicle.com/article/will-your-college-survive-the-demographic-cliff
8 Gershenfeld, Neil, *Fab: The Coming Revolution on Your Desktop - From Personal Computers to Personal Fabrication* (New York: Basic Books, 2007); Toffler, Alvin, *The Third Wave* (New York: Morrow, 1980).
9 Nicholson, Kendall A, "Where are the women? Measuring progress on gender in architecture." *Association of Collegiate Schools of Architecture* (2020).
10 Innovate at Iowa State Website. https://innovate.iastate.edu/. Accessed 29 August 2022.
11 Precarity Lab, *Technoprecarious* (London: Goldsmiths Press, 2020).
12 Haraway, Donna, "Situated knowledges: The science question in feminism and the privilege of partial perspective." In *Feminist Theory Reader*, pp. 303–310 (London: Routledge, 2020).
13 Ibid.
14 Innovate at Iowa State Website. https://innovate.iastate.edu/. Accessed 29 August 2022.
15 AHA Sends Letter to ISU Urging Reconsideration of Planned Budget Cuts (March 2022). https://www.historians.org/news-and-advocacy/aha-advocacy/aha-letter-to-iowa-state-university-urging-reconsideration-of-planned-budget-cuts-(march-2022).
16 See Smith, T'ai, *Bauhaus Weaving Theory: From Feminine Craft to Mode of Design* (Minneapolis: U of Minnesota Press, 2014) and Forlano, Laura, Molly Wright Steenson, and Mike Ananny, eds. *Bauhaus Futures* (Cambridge, MA: MIT Press, 2019).
17 Ibid.
18 Computer Science 18% Female, 82% Male (Source NSF, 2018); Registered Architects 19% Women, 81% Men (Source AIA, 2020) and Construction Workers 10% Female, 90% Male (Source: Bureau of Labor Statistics, 2020).
19 See Crenshaw, Kimberle, "Demarginalizing the Intersection of Race and Sex: A Black Feminist Critique of Antidiscrimination Doctrine, Feminist Theory and Antiracist Politics," University of Chicago Legal Forum: Vol. 1989: Iss. 1, Article 8. Available at: http://chicagounbound.uchicago.edu/uclf/vol1989/iss1/8 and Collins, Patricia Hill. *Intersectionality as critical social theory*. Duke University Press, 2019.

20 Gay, Roxane, *Bad Feminist: Essays* (Munich: btb Verlag, 2019).
21 MIT Design and Computation and https://architecture.mit.edu/computation. Accessed 5 December 2022.
22 papers.cumincad.org CumInCAD is a cumulative index of publications about computer-aided architectural design. It includes bibliographic information about over 12,300 records from journals and conferences such as Association for Computer Aided Design in Architecture (ACADIA), The Arab Society for Computation in Architecture, Art and Design (ASCAAD), Association for Computer-Aided Architectural Design Research in Asia (CAADRIA), Education and research in Computer Aided Architectural Design in Europe (eCAADe), Sociedad Iberoamericana de Gráfica Digital (SiGraDi), Computer-aided architectural design (CAAD) futures, (DDSS), and others. All papers include full abstracts. Full texts, in PDF, of some 9,600 papers are also available. CumInCAD was designed and created by Prof. Bob Martens, Vienna University of Technology (Austria), and Prof. Ziga Turk, University of Ljubljana (Slovenia), in 1998.
23 ISU Department of Architecture faculty and collaborators Nick Senske (2016-present), Leslie Forehand (2016–2019).
24 2017. acadia.org Disciplines & Disruptions at Massachusetts Institute of Technology the 37th conference of the Association for Computer-Aided Design in Architecture. ACADIA was formed for the purpose of facilitating communication and critical thinking regarding the use of computers in architecture, planning, and building science. The organization is committed to the research and development of computational methods that enhance design creativity, rather than simply production, and that aim at contributing to the construction of humane physical environments. A particular focus is education and the software, hardware, and pedagogy involved in education.
25 Doyle, Shelby; Forehand, Leslie; Senske, Nick. (2017) Computational Feminism: Searching for Cyborgs ACADIA 2017: DISCIPLINES & DISRUPTION [Proceedings of the 37th Annual Conference of the Association for Computer Aided Design in Architecture (ACADIA) ISBN 978-0-692-96506-1] Cambridge, MA 2–4 November, 2017), pp. 232–237. https://doi.org/10.52842/conf.acadia.2017.232.
26 Rendell, Jane, "Only resist: A feminist approach to critical spatial practice." *Architectural Review* 19 (2018). See also Sexuality & Space by Jennifer Bloomer and Beatriz Colomina.
27 Bridle, James, *New Dark Age: Technology and the End of the Future* (New York: Verso Books, 2018).
28 Carpo, Mario, ed, *The Digital Turn in Architecture 1992-2012* (Hoboken, NJ: John Wiley & Sons, 2012).
29 Carpo, Mario, "The second digital turn." *The Second Digital Turn* (2017).
30 Behar, Katherine, ed., *Object-oriented Feminism* (Minneapolis: U of Minnesota Press, 2016).
31 Ibid.
32 Doyle, Shelby Elizabeth, and Nick S. Senske, "Digital Provenance and Material Metadata: Attribution and co-authorship in the age of artificial intelligence." *International Journal of Architectural Computing* 16(4): 271–280. Doi:10.1177/1478077118800887.
33 Haraway, Donna J, *Staying with the Trouble: Making Kin in the Chthulucene* (Durham, NC: Duke University Press, 2016).
34 Haraway paraphrases the work of Marilyn Strathern as described in a lecture at Arizona State's Institute for Humanities Research. vimeo.com/62081248. Accessed 20 December 2022.
35 Fisher, Berenice, and Joan Tronto, "Toward a feminist theory of caring." *Circles of Care: Work and Identity in Women's Lives* (1990): 35–62.
36 Bridle, James, *New Dark Age: Technology and the End of the Future* (New York: Verso Books, 2018).
37 Benjamin, Ruha, "Race after technology: Abolitionist tools for the new Jim code." *Social Forces* (2019). See also McKittrick, Katherine, *Dear Science and Other Stories* (Durham, NC: Duke University Press, 2020).
38 Doyle, Shelby Elizabeth, and Nick S. Senske, "Women's work: Attributing future histories of the digital in architecture." *The Plan Journal* 4, no. 2 (2019) [GENDER MATTERS]: 307–326.
39 Doyle, Shelby Elizabeth, and Nick S. Senske, "SOM's Computer Group: Narratives of women in early architectural computing." *International Journal of Architectural Computing* 19, no. 3 (2021): 213–225.
40 Bridle, James, *New Dark Age: Technology and the End of the Future* (New York: Verso Books, 2018).
41 de La Bellacasa, Maria Puig, *Matters of Care: Speculative Ethics in More than Human Worlds* (Minneapolis: U of Minnesota Press, 2017).

Bibliography

AHA Sends Letter to Iowa State University Urging Reconsideration of Planned Budget Cuts (March 2022). https://www.historians.org/news-and-advocacy/aha-advocacy/aha-letter-to-iowa-state-university-urging-reconsideration-of-planned-budget-cuts-(march-2022).

Alvin Toffler. *The Third Wave*. New York: Morrow, 1980.

Berenice Fisher and Joan Tronto. "Toward a feminist theory of caring" in *Circles of Care: Work and Identity in Women's Lives,* eds. Emily K. Abel and Margaret K. Nelson, pp. 35–62. New York: SUNY Press, 1990.

Donna Haraway. "Situated knowledges: The science question in feminism and the privilege of partial perspective." In *Feminist Theory Reader*, pp. 303–310. London: Routledge, 2020.

Donna J. Haraway. *Staying with the Trouble: Making Kin in the Chthulucene*. Durham, NC: Duke University Press, 2016.

Elizabeth Fisher. "The Carrier Bag Theory of Evolution," in *Worm's Creation: Sexual Evolution and the Shaping of Society*. New York: McGraw-Hill, 1980.

Neil Gershenfeld. *Fab: The Coming Revolution on Your Desktop - From Personal Computers to Personal Fabrication*. New York: Basic Books, 2007.

Innovate at Iowa State Website. https://innovate.iastate.edu/. Accessed 29 August 2022.

James Bridle. *New Dark Age: Technology and the End of the Future*. New York: Verso Books, 2018.

Jane Rendell. "Only resist: A feminist approach to critical spatial practice." *Architectural Review* 19 (2018). https://www.architectural-review.com/essays/only-resist-a-feminist-approach-to-critical-spatial-practice

Jon Boeckenstedt. "Will your college survive the demographic cliff." *Chronicle of Higher Education*. https://www. chronicle. com/article/will-your-college-survive-the-demographic-cliff (2022).

Katherine Behar, ed. *Object-oriented Feminism*. Minneapolis: U of Minnesota Press, 2016.

Katherine McKittrick. *Dear Science and Other Stories*. Durham, NC: Duke University Press, 2020.

Kendall A. Nicholson. "Where are the women? Measuring progress on gender in architecture." *Association of Collegiate Schools of Architecture* (2020). https://www.acsa-arch.org/resource/where-are-the-women-measuring-progress-on-gender-in-architecture/

Kimberle Crenshaw. "Demarginalizing the intersection of race and sex: A Black feminist critique of antidiscrimination doctrine, feminist theory and antiracist politics." University of Chicago Legal Forum 1989(1): Article 8.

Laura Forlano, Molly Wright Steenson, and Mike Ananny, eds. *Bauhaus Futures*. Cambridge, MA: MIT Press, 2019.

Maria Puig de La Bellacasa. *Matters of Care: Speculative Ethics in More than Human Worlds*. Minneapolis: U of Minnesota Press, 2017.

Mario Carpo, ed. *The Digital Turn in Architecture 1992-2012*. Hoboken, NJ: John Wiley & Sons, 2012.

Mario Carpo. *The Second Digital Turn*. Cambridge, MA: MIT Press, 2017.

MIT Design and Computation. https://architecture.mit.edu/computation. Accessed 5 December 2022.

Patricia Hill Collins. *Intersectionality as Critical Social Theory*. Durham, NC: Duke University Press, 2019.

Precarity Lab. *Technoprecarious*. London: Goldsmiths Press, 2020.

Rob Stevens. "Presidential initiative sets sights on high-impact faculty hires", 2013. https://www.inside.iastate.edu/article/2013/09/12/hire. Accessed 5 December 2022.

Roxane Gay. *Bad Feminist: Essays*. Munich: btb Verlag, 2019.

Ruha Benjamin. *Race After Technology: Abolitionist Tools for the New Jim Code*. New York: Polity Press, 2019.

Shelby Doyle, Leslie Forehand, and Nick Senske. "Computational Feminism: Searching for Cyborgs." ACADIA 2017: DISCIPLINES & DISRUPTION [Proceedings of the 37th Annual Conference of the Association for Computer Aided Design in Architecture (ACADIA) ISBN 978-0-692-96506-1] Cambridge, MA, 2–4 November, 2017), pp. 232–237. https://doi.org/10.52842/conf.acadia.2017.232.

Shelby Elizabeth Doyle and Nick S. Senske. "Digital Provenance and Material Metadata: Attribution and co-authorship in the age of artificial intelligence." *International Journal of Architectural Computing* 16, no. 4: 271–280. Doi:10.1177/1478077118800887.

Shelby Elizabeth Doyle and Nick S. Senske. "Women's work: Attributing future histories of the digital in architecture." *The Plan Journal* 4, no. 2 (2019) [GENDER MATTERS]: 307–326.

T'ai Smith. *Bauhaus Weaving Theory: From Feminine Craft to Mode of Design*. Minneapolis: U of Minnesota Press, 2014.

Ursula Le Guin. *The Carrier Bag Theory of Fiction*. Barcelona: Terra Ignota, 1986.

Ursula Le Guin. "A Rant About 'Technology,'" 2004. http://ursualkleguinarchive.com/Note-Technology.html.

Situated Computations

An interview with Vernelle A. A. Noel

GALO CANIZARES: As an introduction, let's speak a little bit about your entry into the world of computational design and digital fabrication. I think the way you have described your work is quite unique. For example, you say that your goal is to build new expressions, new tools, and methodologies to explore the social, cultural, and political aspects of making technologies, computational design, and emerging technologies. This suggests that there's a lot of sides to your work and academic interests. How did you get here from your education and training?

VERNELLE A.A. NOEL: I initially studied civil engineering at home in Trinidad & Tobago. I always wanted to study architecture, but unfortunately, it was not immediately accessible to me, so I studied civil engineering instead. One of my teachers recommended that I study civil engineering since it would give me more options in the construction industry. So that's what I did. I worked as an engineering surveyor, architectural technician, and engineering technician while working on my diploma. I worked as an architectural technician for several years, including my own private projects, and thought that that would quell my desire to be an architect, but it didn't. I knew there was so much more I wanted to learn. It was my brother who seriously suggested that I pursue my studies in the US. I could not see that as an option because I knew my family could not afford it. I am making a long story short here, but I eventually got the opportunity to join the B.Arch

DOI: 10.4324/9781003296522-20

program at Howard University in Washington, DC. While at Howard, I was still practicing and working. After my graduation, I chose to work with Burt Hill (now Stantec) in DC, then after a year, moved to Gujarat, India to help establish their new office there as a BIM Manager and Project Architect. When the economy crashed in 2009, I went back to Trinidad and Tobago. I did some consulting for a firm there, then a friend put MIT on my radar. The computation program looked quite interesting as I was working in BIM at the time and that's what I thought I would focus on, but it became something different.

Digital fabrication was quite new to me. While at Massachusetts Institute of Technology (MIT), I took a few classes in the Media Lab, and a formative one for me was Neil Gershenfeld's famous "How to Make (almost) Anything" class. In this course, we explored different fabrication techniques and combined engineering, design, and programming. The making part was easier for me than programming – which was a whole new world to me. A second was working with Larry Sass who also does research in digital fabrication. These two experiences were formative in my understanding of computational design and digital fabrication (Figure 17.1).

Figure 17.1 Vernelle A. A. Noel conducting a wire-bending workshop at a high school in Trinidad & Tobago (2017).

GC: Now you are the director of the Situated Computations + Design Lab at Georgia Tech. What kind of projects and research do you engage with in that role?

VN: Building and reconfiguring computational design frameworks, methodologies, tools and more that critically engage in hi and low-tech ways of thinking and design, and how our work actually touches the ground and those we claim our work is for. Architecture is much, much broader than the profession – it is a way of seeing the world, making the world, and interacting with the world. I want to broaden and expand concepts, definitions, practices, participants and more in our field and ground it in ways that question: who participates or not in computation, what we call computation (or not), how we theorize and conceptualize computation, and how we create space for those currently absent from the field.

GC: In your description of the lab there are a lot of mentions of interdisciplinarity. This goes along with your desire to go beyond (or pop) the disciplinary bubble. Can you elaborate a little bit on how you approach interdisciplinarity? Do you collaborate with others from different disciplines or do you yourself learn different techniques and methods from other disciplines?

VN: My perspectives and experiences has been honed by many things: engineering, architecture, art, traditions, etc. My way of seeing the world is multiple and varied. I am also inspired and excited by the ways other fields do things. My interdisciplinarity comes from my experiences and an interest in cool things other people are doing in other fields. Again, the goal is to get out of the bubble. Multiple perspectives are necessary to address the issues and problems that press us today. It's also important because it shapes how outsiders understand what we bring to the table.

My interdisciplinary positions are based on methods, concepts, theories. Whether they are representational methods or quantitative/qualitative methods from STS (Science and Technology Studies), I try to create bridges between traditional and high-tech practices. I also teach an interdisciplinary studio which I enjoy very much. One year, I collaborated with faculty and students in computational media. Another year, I chose to collaborate with a faculty member in aerospace engineering whose expertise is also in lightweight structures. I'm also working with an anthropologist on another project. I love forming relationships with people from different areas of expertise. It exposes me to not only different research and teaching topics and methods but also to different perspectives and values. Apart from expanding my network of collaborators, I try to instill in my students the notion that the world is best experienced in an interdisciplinary way. Once you get out of your silo, you gain so much more richness from the world. It also allows students to learn different ways of communicating and understanding different contexts. This is the most important part of stepping into roles that are different and unfamiliar (Figure 17.2).

GC: If we look at the work that you've been doing with the Trinidad Carnival, that seems to be a good example of you wearing a lot of different hats. You talk about your work as part ethnography and part historical research on the history of Trinidad and

Figure 17.2 The Infinite Line Active-bending pavilion – designed based on material behavior and fabricated using techniques from wire-bending on exhibition at International association for shell and spatial structures (IASS) in Barcelona, Spain (2019).

Tobago. Then there's obviously computational design, which is your expertise. How do you switch between these modes when you're doing design research? Is there an organizational technique that you require to do this?

VN: I wouldn't call it a switching, but there's definitely a lot of contexts that must be unpacked before engaging in design; a lay of the land that I feel that I need to get before thinking of design possibilities. I never go into a project thinking: "Oh I want to try out this cool computational design." No. The work is always grounded in a site. I must really know the site; I must know the people; I must know their values. There's a reality that must be uncovered first, and then have discussions around research questions, project ideas, and computational design that are based on that reality. As designers we're uniquely used to jumping around and wearing different hats, so I don't think this is anything new. We must always investigate the whys and the why nots. And this takes time. It is possible to build something in a semester but developing relationships and trust with the people takes more time. It is important to be aware of the time, trust, and all that informs the potential of a project. It's never as simple as just writing a script and executing it (Figure 17.3).

GC: It seems that this approach forms the backbone of your idea of Situated Computations. Could you describe a little bit what that term means and maybe how

Figure 17.3 "From Dusk to Dawn" – designed and constructed using wire-bending materials, techniques, and steps extended from the Bailey-Derek Grammar (2019).

you arrived at this concept? Was it something that you felt was lacking in design disciplines? What is the ongoing history of Situated?

VN: Situated Computation comes from a reflection of my own work in craft, technology studies, and digital fabrication. It pulls from digital media studies and the contexts in which I work. I have essentially sat back and reflected on my research projects, with an emphasis on aspects that have worked well. The purpose of Situated Computation is, again, a way for us to ground our field and make visible the faces, the people, and labors in our work. Aspects that are often made invisible when we focus on artifacts or technologies or the objects we make. We rarely see people in the work of digital fabrication. You might see someone standing by their work, but we don't really see people *in* the work. We see things, we see technologies, but technologies don't exist without people. The concept also addresses the apolitical attitude of many design professionals. If politics is about power relations, then there's nothing that is apolitical, especially in computational design and digital fabrication.

Situated Computation also questions what "technology" might mean in other contexts. In the context of Caribbean practices, for example, there are no big grants or millionaires funding your projects. This is not how we make projects. We are innovative and we make diamonds out of the simplest things. So, the question is how do we talk about computation in the context of other abundances and values that may not be solely technological or economic? One answer might be to look at

Figure 17.4 Handout showing a snippet of the Bailey-Derek Grammar being used by participants during wire-bending workshop (2017).

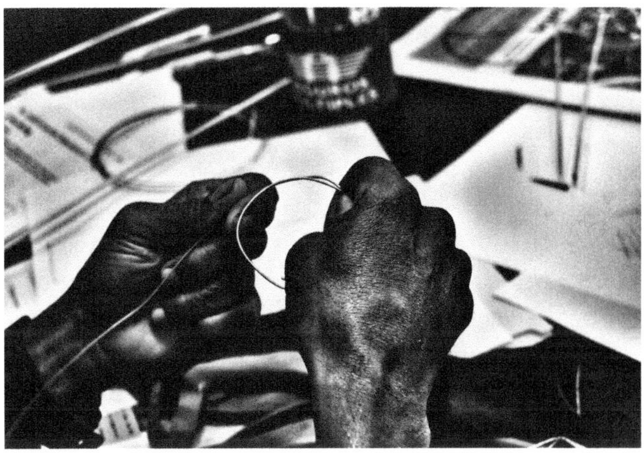

Figure 17.5 Workshop participant engaging with wire during wire-bending workshop (2017).

traditions. In the Caribbean, we innovate with what we have; we "bricolage." This could be considered a kind of computation, a way of making a whole out of parts according to some rules or logics. Thinking in this way, with an expanded definition of technology, is how we can create connections with real people in the real world. Getting feedback from the people out in the world is also important. This is another way we break the bubble. We should ask what is computation in the streets. How can we compute without huge computing resources? We should be imaginative enough to be able to think outside the notion that technology and digital fabrication can only be done with expensive machines (Figures 17.4 and 17.5).

GC: I love that phrase, "computation in the streets." It's a really powerful statement. Let's go a bit deeper into the Situated Computation concept. In one of your papers, you outlined 8 points for a *Situated Computation approach*. It reads almost like a manifesto and follows the tradition of design theorists outlining bullet points for specific principles that should be followed. Can you talk a bit more about those principles and why it was important for you to outline those in such a way?

VN: The principles of Situated Computation are: (1) be built on an ethnographic study, (2) be situated, (3) build on existing skills and knowledges, (4) facilitate sensory, perceptual interaction, and physical manipulation of materials and tools, (5) build communities, not individual isolation, (6) cater to experts and non-experts in CAD and computer programming, (7) do not require large amount of computational power and infrastructure, and (8) narrative.

Number 1 addresses what I mentioned earlier regarding context. We must be aware of what is happening on the ground, and we must disregard any assumptions we might have. This is where ethnographic studies come in. Ethnographic methods allow us to unpack the complexities of a site: its politics, its social values, its culture, and its environment. All of these elements inform our relationship to technology, and so they must be taken into account when starting a project somewhere. Designers should not be scared about doing ethnographic work. It might feel a bit outside of design work, but it is crucial for gaining a solid understanding of a site. Number 2 emphasizes this: we, as designers and researchers, need to be on the ground asking questions. We need to be listening to people and understanding exactly what is happening in these communities. Number 3 is all about designing for specific contexts. We shouldn't design things that people within a particular context don't have the skills or resources to use. We should position ourselves more as learners than experts and become familiar with the skillsets of the people we are working with rather than impose one way of doing things. If we learn what the existing skills are, we may be able to support or augment them. In order for computation to happen, first we must understand what kinds of knowledges exist there. Can computation emerge out of dance, singing, or handcraft? I think so. Number 4 addresses materiality. Situated Computation resists structures of separation and purity; it aims to reveal sensory effects and material interactions beyond the abstract. The tools that I make foster active engagement and intermingle with sense, bodies, and materials. Number 5 reinforces what I've already mentioned about community. It is critical to build communities around technologies to foster participation and shared learning. Number 6 is also related to community building. We should prioritize the mingling of experts and non-experts so that fruitful collaborations can emerge. We are all experts and non-experts. Both can complement one another and arrive at totally unexpected outcomes. Number 7 asks computational designers to think beyond the resources of Western contexts. Not every client has millions of dollars to fund a computational project. Nations have lost economic power through wars, colonialism, slavery, and other forms of domination which persist. Designers should be mindful of the technological infrastructures of a site and region, especially when working on a digital fabrication project. The last point, Number 8 is simply "narrative." Situated Computation tells the stories of the people involved in our world, their histories, and their innovations. There's a saying that goes something like "until the lion learns to write, history will always be the story of the hunter as a hero." We know all too well that there are many people whose stories are never told or even erased. We should be giving a voice to marginalized and disenfranchised groups because their knowledges and practices are beneficial today.

GC: Storytelling is a type of knowledge, too.

VN: Yes, oral histories are very important in African cultures – it is how knowledge became codified. You have carvings, illustrations, songs, and other ways beyond words and text that people have theorized on life and explained phenomena. We should be actively making space for these multiple ways of expression.

GC: Maybe we can talk a bit about aesthetics and form. Many of your projects involve lines and splines. Obviously, the wire bending projects follow that line of thinking. It's almost like your interest in computational geometry is centered around: what is the most one can do with lines and curves. It reminds me of old ship-building and aeronautical techniques where draftspeople would use physical wood splines to draw curves. It's no secret that before a spline was a computational line, it was bendy piece of wood. Are you connecting to that history in some way? Why are you so fascinated with lines and curves?

VN: I really love lines. I think they're so malleable, they're so present in real life and in digital life. I often just draw lines as doodles and sketches. Of course, there's a problem with that because in architecture we need planes and other solid geometries. So, there's a wrestling that I have to engage with and figure out when it comes to building architecture. But I don't know why I'm so obsessed with lines, I just am. Maybe it comes from my focus on making. Meaning that I'm able to see the line as a physical thing before it's a digital thing. Whether a line springs, whether it dances, whether it bends, whether it is a hard straight line, I look at it as a material object. What I genuinely love about lines is their interaction with light and shadow. How they look when light is cast on them. That's the most exciting part for me in making these line structures. There are also so many other kinds of lines: the lines of the body, the lines that trace movement, lines that are woven. There's so much that can be unpacked with just lines (Figure 17.6)!

GC: It's interesting because it feels like it's really leveraging some very fundamental components of software, which is the spline itself, a basic element which is typically used for constructing something else. But many of the projects that I've seen come out of your work and your work with students at workshops and conferences physicalize those lines in some way. It's almost as if the negotiation between physical and digital is at its most tenuous in these line-based projects. Can you talk about that?

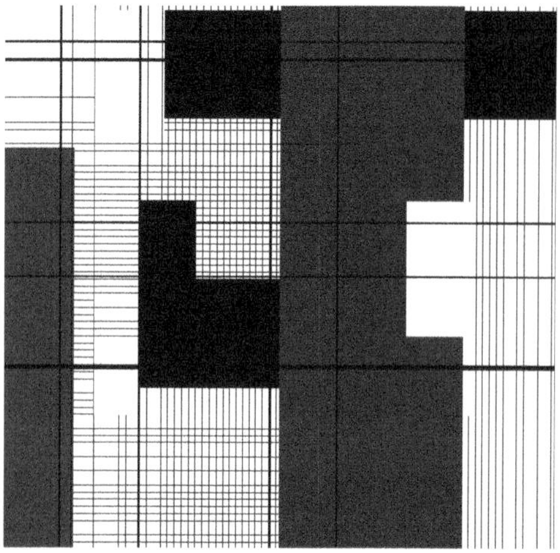

Figure 17.6 Web-based drawing application "Naladrian" plays with shapes and lines (2021).

Figure 17.7 Wire-bending structure for "the boat" for Mathemalchemy.org – 24 mathematical artists and artistic mathematicians (2021).

VN: My exploration of lines whether in light paintings, sketches, or sculpture, are embodied things that I explore physically. They are real things. I draw them by hand, I bend them by hand, I move my body around them. They're physical things that also become digital things, and digital things that become physical things (Figures 17.7 and 17.8).

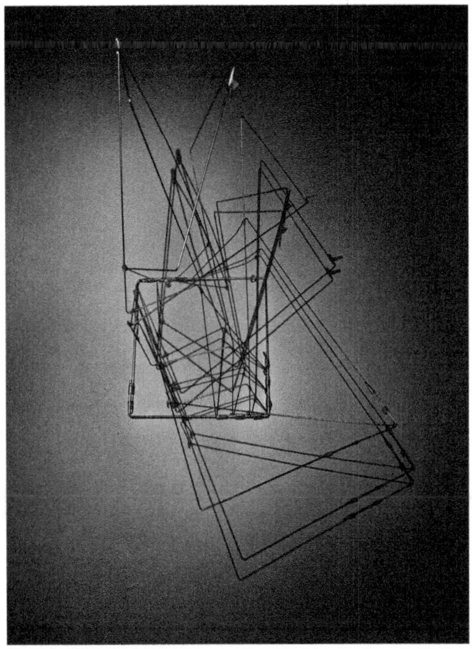

Figure 17.8 Wire-bending sculpture – "Explosion," (2021).

GC: That negotiation between physical and digital also seems to be centered on an idea of craft. Craft is a curious word in architecture because it is something we often avoid talking about. I think there's a common misconception that craft is different from technology or that traditional craft is at odds with high design or high-tech design. But in your work, you talk a lot about craft, and you engage traditional craft in conjunction with computation. The term craft is also gendered at times, associated with the feminine. That may be one reason why it has been isolated in discussions of technological disciplines. Can you talk a bit about why you're interested in craft?

VN: My interest in craft came about during my graduate studies. I knew I wanted to do research in/ about Carnival in Trinidad and Tobago, and in the early stages of research, I found out about a particular craft that was dying: wire bending. My ethnographic work involved interviews with wire-benders and other designers and makers in Carnival, observations, etc. I began looking at the tectonics of it and its experts. There was a strong sense of humanity in this craft. Of course, I also had to find some way to bring computation into it, as this was my thesis. I used computation to document and make the largely undocumented practice of wire bending more explicit. There was a gender component to it as well. Wire bending is traditionally a male-dominated practice. My goal was to use the computational tools I developed to bring women and children into the conversation. It became a project about accessibility and access to knowledge, all revolving around this unique craft (Figure 17.9).

Figure 17.9 "The Sail" – designed from analysis and parametrization of traditional characters in the Trinidad & Tobago Carnival and digital fabrication methods (2016).

GC: Is this where you started to work with shape grammars (the theoretical computational design technique invented by George Stiny at MIT), which is in essence a kind of non-computer-based computation? How does this fit into the project?

VN: Shape grammars fit into the project because of its accessibility. I could use pen, paper, and pencils – the same tools as wire benders – to have a discussion about computation and wire bending. They can quickly assess whether something works or doesn't in a very simple way. Shape grammars are very simple, abstract rulesets and do not really require advanced computation. It made it easy, accessible, and cheap to produce and iterate through designs. The rules that were developed could be communicated to the community without any software or hardware. It was exciting to be able to work at both extremes: the low-tech and the high-tech. It helped me realize that they are very similar ways of working, and that there is much richness to uncover when we map one on top of the other.

GC: Have you encountered any limits to the approach of Situated Computation? I'm thinking that there's always more that can be done. There's more to learn about the site, more to learn about the people, more about traditions, etc. And projects always have constraints whether its time or budget. Or both. How do you make value judgements about what should be part of the project and what you are not going to be able to include?

VN: Time and resources are the biggest hurdles. With more people, we could get more things done. There are experiments and workshops that I want to conduct in Trinidad & Tobago, and the entire Caribbean for that matter, but between a pandemic and the cost to run workshops, to ship technology, and travel there, I'm very constrained. These are the realities of research that are at times assumed, unknown, or just ignored. In terms of parts of a project that could be left out, it depends on the project. But based on the principles I mentioned earlier, there are some components that must always be there such as ethnographic research and conversations with people. My rule of thumb is that the ground and the people are what matter the most.

It is a tricky realm to navigate because as a tenure-track academic, there are parts of research that cannot be ignored. But my community isn't comprised of just people reading academic papers. I want people outside architecture and academia to know what I'm doing. I want the person on the street to be able to critique my work and for us to have a conversation about it. Any limits that I have encountered typically revolve around this negotiation between insider and outsider knowledge.

GC: Would you say that if a project had a lot of constraints, priority would always be given to the site and the people versus the technology?

VN: Yes, but, again, we should broaden how we think about technology. The shape grammar technique is a technology, too, for example. It's not a digital technology but it is a form of computation that can be done with very limited resources. We can still make progress by using these types of technologies that require little infrastructure. My point is that our research should make space for those who may not necessarily have hi-tech resources but are still interested in computational things. Like George Stiny and Terry Knight at MIT. They're not pursuing AI or advanced 3D printing. Yet, their theories of shape grammars allow them to reflect critically on geometry and fabrication at an abstract level.

Beyond Default Thinking

A Conversation with Dr. Behnaz Farahi, Dr. Madeline Gannon, and Mollie Claypool

GALO CANIZARES: The notions of critical making and reflecting on design's subjects are quite provocative. We've been asking a lot of the practitioners and participants in this book to reflect on some questions that allow you to recalibrate and rehome your practices. Are there any other questions or concepts that you go back to in terms of origins? We're using the term homing as a metaphor for working within this type of practice. Do you have concepts that you go back to or maybe ideas that you were inspired by as you dove deeper into the world of digital fabrication?

DR. BEHNAZ FARAHI: The world of computational design has tended to overemphasize the importance of technology, often leading to a focus on producing aesthetically pleasing forms through digital fabrication and design. This is particularly evident in the fields of computational design and architecture, where there is often a fixation on creating novel, intriguing forms that do not extend beyond formal experimentation. In recent years, there has been a shift in how we approach computational design. It's no longer enough to focus solely on technology; instead, we must consider how it can be used to address broader societal issues. This is where criticality comes in; it challenges the use of technology and raises questions such as "Why was this made?" and "Who is the intended audience?" (Figure 18.1).

Initially, my practice centered on utilizing tools for digital fabrication, computational design, and robotic fabrication. However, in recent years, I've shifted my

 DOI: 10.4324/9781003296522-21

Figure 18.1 Mesolite: An emotive display, Behnaz Farahi, 2018.

focus to explore how computational design can address sociocultural and political issues. I believe that solely exploring form results in a limited audience, and as architecture is already an insular field, it's crucial to broaden its scope and bring in diverse perspectives. Hence, my aim is to engage in conversations with non-architects and non-designers that could enable us to have a greater impact and reach a larger audience.

MOLLIE CLAYPOOL: I think you're right, Behnaz. It's important to expand the set of stakeholders that can interact with our work. For me, it comes down to tangibility. I like to reflect on what the infrastructure is that we're making or building that enables work to become tangible. One of the examples I come back to is the cooperative—an organizational structure that is run and owned by its members but allows individuals to work towards their individual goals, outcomes, and benefits rather than simply sharing expertise towards achieving the same goal. It allows shared priorities without requiring consensus, like collaboration does. I believe collaboration can make it very difficult for true intersectionality in design. I'm interested in cooperating together (Figure 18.2).

GC: That's interesting because I think we talk a lot a lot about collaboration, especially in academic contexts. We talk about interdisciplinary collaborations and applying for grants together with engineers and entrepreneurs, etc, but rarely in the form of consensus. I haven't thought about it in that way before. Behnaz, you work with a lot of people in other fields as well. Your work crosses into fashion and wearables. That scale of work is captivating because I see it becoming a bit more popular or maybe it's just that the technology is advancing, and people are becoming more aware of this sort of smaller scale digital fabrication. How would you characterize the kinds of collaboration or cooperation that you're engaged in?

Figure 18.2 House Block, AUAR, 2021.

BF: It was an intriguing journey for me, as my background is in architecture, not fashion. Despite this, many people mistakenly label me as a fashion designer. My fascination with the fashion industry, however, stemmed from the limitations of digital fabrication. While I was an artist-in-residence at Autodesk's Pier 9 workshop in San Francisco, I utilized Stratasys' multi-material 3D printing technology, which had size restrictions that posed a challenge. I was determined to utilize algorithmic design tools, compositions of matter, and various geometrical features to create intricate designs that could only be produced using digital fabrication. This led me to consider the body as a scale for fabrication. If I had access to larger robots and materials, I may have designed a pavilion or a large-scale installation instead. However, the human scale presented me with a new set of tools and ideas to explore. I began to contemplate the human body, human perception, and the criticality of the female body and its relationship to occupying space. This world of fashion and wearables has piqued my interest in exciting new ways.

GC: That's a nice way of using the limitations or constraints that are afforded by digital fabrication technologies to your advantage. It's also one of the most common ways that people find a site or a project or an area within this kind of subfield. Makers tend to find potentials in the constraints of a machine. So in a way, the machine dictates the scale in which they operate. And then there's the other perspective, which is that digital fabrication sometimes is seen as strictly a kind of prototyping arena. Often the line of thinking is that an innovative pavilion is a test for large-scale construction. This suggests that there should always be a way of scaling up of techniques, tools, or processes. What I really appreciate in your answer, was that you found a whole world that exists at the scale that you were comfortable with, and then you just are tapping into that, and focusing more on expanding into that world, rather than saying something like *this jacket can become a building*.

DR. MADELINE GANNON: Or seeing this object like a facade prototype! It's a huge bias that we architects have: *how do we take this cool thing and shove it into a building?* It diminishes our expertise in a way because we know so much more than just buildings. Architects understand how the body moves through space and the social and technical implications of that. Being able to work at a variety of scales is another core skill. I met Behnaz through an artist residency when we were both working on 3d printing for the body. Her work was so incredible; she was making an interactive garment *and* an interactive ceiling. What struck me was that the garment had the same number of actuators as the entire ceiling of a room. The impact of her garment was so much more visceral and powerful–I think because the dynamism of its movement was over a much smaller area than its architectural equivalent. One challenge with building interactivity into architecture is that when you scale up, not only does it cost more, but its impact–what it communicates to inhabitants–isn't as palpable nor potent as when you relate it to the human body (Figure 18.3).

BF: Thank you, Madeline. I mean, it wasn't just like me or Madeline doing this. There are several other architecturally trained individuals doing this; people like Neri Oxman and Niccolo Casas. I was amazed by how many architects in the field are doing body size objects, wearables, or other forms of 3d printing what we call "body architecture." That research at Autodesk Pier 9 led to the AD publication that we did with Neil Leach on 3d Printed Body Architecture, where we

Figure 18.3 MANUS, ATONATON, 2018.

featured for instance, Madeline's work and a lot of other amazing designers and architects that were working at that scale. There was a whole group of architects that were exploring how digital fabrication could be implemented at the scale of the body.

GC: The body also implies a whole other set of relations. It's not just spatial. There's also identity and expression and the politics that go into collective bodies versus singular bodies, the male body, the female body, etc. It gets quite complicated as soon as you start to talk about the body and I think that's something that has been overlooked in digital fabrication. As editors, we're interested in what distinguishes the current discourse on the body from previous discourses about bodies in space, for example. How does that relate to the emerging approaches of technology? Madeline, for instance, you talk a lot about behavior and interaction. I wonder if you can talk a little bit about your approach regarding the body. Does this impact feedback between machines and the body?

MG: Just to go back to origins for a second: my wandering from architecture into robotics really happened through computer-aided design manufacturing. Long story short, I came out of school during the great recession, when there were no jobs. I decided it might be better to specialize in something, and so I stayed in school. I really wanted to be a curator, but I didn't get into any of those programs. I got into a program at Carnegie Mellon, which just so happened to be super technical. Kicking and screaming, I learned how to program, and I started to figure out how to talk to machines that can make things.

My goal was to make tools that would be more intuitive, more interactive, and have direct feedback–tools that could do things that big, mass-market CAD tools couldn't. This led me to focus on interactive design and fabrication tools. It just so happened that the lab where I was working had 3d printers, laser cutters, CNC routers, and a big giant robot sitting there doing nothing, because no one knew how to use it. It was a natural progression for me to think about interaction–about intuitive, continuous feedback–with that six-axis machine (the robot) instead of the 2.5- or three-axis machines (like 3d printers or CNC routers).

With my interactive installations, I've been able to take my robots out of the lab and into public environments, and it's been incredible to learn from the feedback of thousands of people experiencing them. I can guess at how people may react, but nothing beats directly observing someone's instinctive response and body language as a one-ton robot comes screaming at them. I try to incorporate these observations in the next generations. What I've found is that something strange happens as soon as you get a robot to move a certain way—it stops being a piece of industrial equipment and starts to seem like a lifelike creature. I am endlessly curious about how our brains are hard-wired to seek connection and personality in inanimate objects, and I'm so excited for a future when we can enchant our environments to life.

GC: That also shifts the perception of a machine from a useful thing to maybe more useless (in the traditional sense).

MG: Well, here's the thing: the entire world is focusing on useful things, right? But there's *plenty* of space-time in between useful tasks where robots are completely useless—and that area is completely undiscovered and unexplored. Delivery robots are a good example: these machines cross hundreds of people on their journey from pick-up to delivery. A delivery robot is simply doing its task, but its mere presence in the world can impact people in different ways. Since architects and designers know a lot about interactions between bodies, space, and time, our training can offer real value to the robotics and technology communities.

GC: I especially appreciate that you're using industrial robots because these have a very loaded history. They come from the assembly line and are part of the complex history of automation. Maybe this brings us to Mollie's work on automated construction. How do you see the kind of behaviors—or the lenses through which Madeline is looking at automated tools—fitting into your ideas of automating labor and construction processes (Figure 18.4)?

MC: We've done a lot of work around community engagement. One of the ways that we like to engage people is by bringing in forms of automation that would otherwise be hidden away from the general public in large factories into places that are more traditional and manual in their practice. For example, we did work in a bricklaying school,

Figure 18.4 House Block, AUAR, 2021.

we've also done work with carpenters. Something that has been amazing to witness is the first moment that a bricklayer sees a robot pick up a brick—in our case, one of AUAR's LEGO-like building blocks—and seeing them get excited because they realize that the backbreaking work that they've done for the last 30 years can be done differently now. We also go out and talk to contractors all the time. Contractors, at least in Europe and North America, are mostly small business contractors. Most of them are one or two people shops. But they contribute something like 9% of GDP in Europe and North America. This means that there's a huge workforce out there doing generally bad work for their bodies, their health, and even their minds. It's a very difficult and dangerous job. This is one of the reasons it has become difficult to recruit younger generations. As a result, there is a massive scarcity of labor in construction. What we're doing is essentially demonstrating that the older generation that can't hand down their business anymore—because their daughter or son doesn't want to go into it—can transform the way that they work and continue with a new kind of practice, while also recruiting younger people into new types of construction jobs enabled by technology. The exciting part is that the technology is becoming increasingly affordable. Whereas five or six years ago you'd be looking at $50–100,000 for an industrial robot, today you can get one for a few thousand. This can enable real change in that industry. There's an amazing interplay between the younger generation—who want to work in jobs that are better—and the older generation that has an amazing wealth of knowledge. Now advanced technology is accessible to both.

GC: That also builds from the perception of technology. A popular notion is that we can now make interesting and futuristic artifacts and that the future is sleek and streamlined. Could you talk a little bit about that perception of technology and how you tap

Figure 18.5 Two Circles, ATONATON, 2023.

into that maybe as a kind of imaginary, either from science fiction or elsewhere? Are there any other perceptions of technology that you want to either pursue or even just debunk as myths?

MG: My biggest gripe with science fiction is that our present is so much more amazing than the futures based on movies from the 60s, which are based on books from the 40s. Where are our imaginations? Where is this generation's imagination that breaks beyond the kind of archetypes and tropes of these tired examples? One of the reasons why I love using existing hardware to do the unexpected is because it demonstrates that this technology—birthed for optimization and efficiency—can be used for communication and care. If you look back at the historical lineage of these tools, you'll find that robotics in the arts and in industry share a similar birth point. The exhibitions *Cybernetic Serendipity* and E.A.T. (Experiments in Art and Technology) happened around the same time as when UNIMATE, the first robot arm, came onto the assembly line. At some point, robots-for-productivity and robots-for-personality bifurcated into two timelines. And we're stuck on the *boring* timeline, where these machines with superhuman powers are just locked away behind cages doing short, repetitive tasks for their whole lives. I want to slip into the *fun* timeline where they are these amazing creatures that coexist with us in ways that enhance our experience of the natural world in new and unimaginable ways (Figure 18.5).

BF: I love the reference to cybernetics here. The notion of 'imagination' and how our field imagines the possible future is quite important. Various design methodologies,

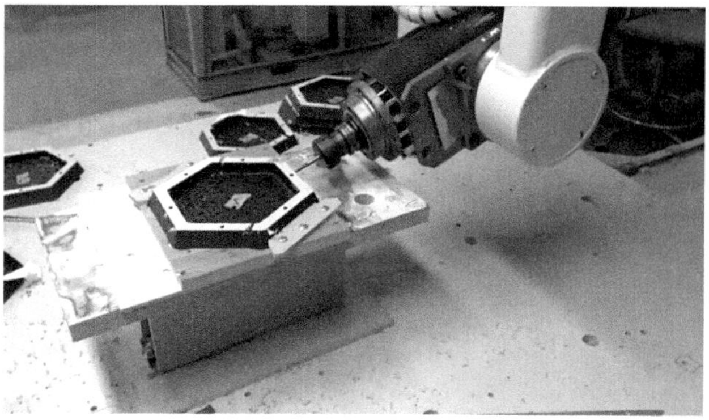

Figure 18.6 5-Axis CNC milling of side holes for modules inter-connections, Behnaz Farahi, 2022.

such as speculative design or world-building projects that pose questions like "what if the world were upside down?", are a constant source of inspiration for me. Whenever I begin a new project, I always start with a similar "what if" question as my entry point. Thinking about possible futures is a thought-provoking approach. Lately, I have been intrigued by other design methodologies like Krzysztof Wodiczko's 'interrogative design'. This approach aims to expose the negative experiences and suffering associated with interacting with designed objects. In this case, critical making is a response to these experiences. Rather than coming up with a shiny and new idea in response to a 'what if' question, this process is much more analytical and includes deep questioning about pains and experiences. This type of imagination is much more nuanced and intricate than typical science fiction tropes and is a refreshing change of pace (Figure 18.6).

MC: When we at AUAR work with groups outside of the discipline, one of the ways that we exercise our imagination is by asking them what's missing? For example, what's missing from your future? Because if you ask people, "what do you need to fix?" they come up with something like, "I need to fix this" or "I would like this." They say "I wish my wall was farther or that I could fit a dresser here." Instead, if you ask, "what's missing from your life?" it opens up this realm of possibility and gives them permission to be creative and act on their imagination. And many times, people have never been asked that question before. They've never been asked, "what's missing from your life?" It's always just "what do we need to fix in your life?" It's a very different approach.

MG: I absolutely love that. I am going to steal it (Figure 18.7)!

MC: It's good, right? If you work in co-creation, the practice of collaborating with other stakeholders to guide the design process, this really should be the approach. It's an asset-based approach. When you say what's missing, you also have to say what's already there. Sometimes people have never been given the chance to recognize

Figure 18.7 House Block, AUAR, 2021.

what's already there, what kinds of resources they have already. By naming them and grounding them, then they can say, "oh, yes, but also I don't have this other amazing thing that I really could use that could help me achieve what I want to be able to achieve in my life or my family's life."

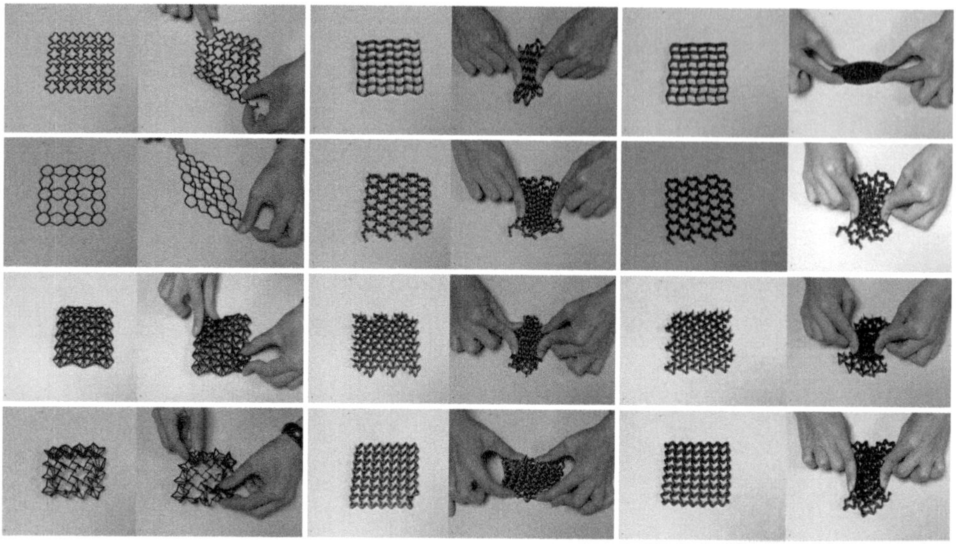

Figure 18.8 Study of auxetic structure using multi material 3D printing, Behnaz Farahi.

MG: It's kind of scary to ask what's missing. When you name what's missing, you're admitting to yourself that you don't have it. Whereas before, maybe you never thought about it and have been coasting or floating through life without it. But knowing what's missing then makes the future more palpable or more actionable. I don't think that question is asked often enough. It's more common to ask what people might need or what they might want.

BF: It's almost like that Henry Ford quote that says "If I had asked people what they wanted, they would have said faster horses." Are we aware of what we want? Or are designers telling people what they should want?

MC: Exactly. We're not telling them what they want. We're not taking that kind of authoritative stance. There was one time when one of my partners on a project went door to door in a neighborhood that we were working in and started asking people what's missing. She knocked on the door of this house that was just a standard two-bedroom semi-detached brick house in suburban Britain. A standard house. The owner opened the door and he said, "you know what's missing from the neighborhood? I'll show you," and he brought her into his kitchen and there was his horse. He said there were no stables for a 20-mile radius and that he didn't drive. So, he didn't have a way of transporting his horse. What he was really saying was that there was no community for him there. He had to hide himself and his horse and couldn't participate in the world around him because his community was not able to be made visible (Figure 18.8).

BF: I think the idea of research, participatory design, and talking to people to understand what is at stake is very important. But sometimes I worry that these layers are so invisible and hidden underneath so much other stuff that even those stakeholders do not exactly know what the problem might be. Sometimes, the problems are just rooted so deep inside that we really need to interrogate to get to the bottom of it. Sometimes people may not know exactly what is missing. Maybe with co-creation, collaboration, or cooperation we can start understanding some of those invisible layers and forces that are often very difficult to understand.

MG: That's where your work does a really good job, Behnaz. Where the architecture discipline fails is in talking to people outside of architecture. We tend to just talk to ourselves. Even something as simple as putting your work on YouTube can break that obscurity. It can open minds and imaginations. When you have a million views on social media, versus a great paper in ACADIA that 20 other people love, you reach a wider audience. And they're both important venues. It's good to connect with the public imagination as well as have your idea vetted by a peer-reviewed publication—that's what will advance the discipline further. But we need more non-standard practices for architecture, in general. More ways that can reach a wider audience and push our collective imaginations forward.

GC: Maybe I can ask a really pointed question then, to all of you: who are your audiences? Or what kinds of audiences do you try to build?

MC: For me, there are two audiences that are interrelated. We work primarily on housing. So, one is people who have access to land. That might be a developer, but in general it's people who own that physical asset or wealth. Then on the other side, another audience is the entire part of the industry that knows how to add value to that physical wealth. That might be contractors, builders, etc. It's a really interesting and difficult problem, because it's a challenge to demonstrate to stakeholder groups that are so driven by a capitalist mindset, that they should do something differently. You have to play into both their imaginations but also have a deep understanding of their mindset and how they work and act. You actually have to understand them as much as possible, which might feel a bit uncomfortable for most architects or designers who have always been taught that they know best. But this is incredibly important because most builders in the world can't access automation. They just can't; it's just impossible because they operate on such low margins in such a cutthroat market that they can't invest in new tools. Developers don't want to invest in automation because they already make profit. They are comfortable in their business model, so they're fine. However, if you can demonstrate that you can do things in a more considerate way, considerate for both the planet and for people, these groups are more excited about change. We say, listen, you know your practice best, but here is a way to gain access to these tools, and you don't have to change too much of your business practices. That tends to get people excited. At that point, you can say that this new process is not only accessible, but it makes the world better for a lot of people from that very granular and individual level of action. Right now, the people we are working with are acting as the first champions of change, the first ripple of many, and we hope that the audience widens more with each ripple (Figure 18.9).

Figure 18.9 MANUS, ATONATON, 2018.

BF: My aspiration for this is to be a source of inspiration for next generations. Being from Iran and being able to do what I have been doing for the last 12 years is not too common. So, I hope to inspire or show people from the Middle East that this career path is possible. I want to reach women from underrepresented communities that have not imagined or thought possible that they could engage with advanced technology, computational tools, or digital fabrication for making forward-thinking projects. Another ideal audience is people that are working on computational projects. I want to engage with people who are asking difficult questions such as "when is it appropriate to use gaze or face-tracking technology?" or "what should we do with artificial intelligence?" These critical audiences are aligned with my values.

MG: I actually ask myself this question on a daily basis: who do I want my audience to be? A lot of my public work was a part of my PhD. My audience then was my dissertation committee. Today, I also have a different audience of people that I need things from. For example, the companies that sponsor me financially with equipment; they need something from me, and I need to give them a return on their investment. Then there are the people that I want to connect with, like the engineers in Silicon Valley. That is a super influential audience where we can illustrate how to tap the full, humanistic potential of certain technologies. For example, I've been working as a robotics researcher for the past five years. And it is so surprising how often the default demo code that ships with the robotic systems focuses on surveillance or mapping—it tends to unfortunately be fairly agnostic to the notions of privacy and agency in space. My hope with tapping into this audience is that I can begin to push at the edges of what industry comes up with and push it more towards the preferable futures that we want instead. I don't think industry is the villain, but there's a big issue with default thinking in these realms—I'm hoping we can change that (Figure 18.10).

Figure 18.10 House Block, AUAR, 2021.

MC: At AUAR, we have developed a building system that has a lot of redundancy ingrained into a lot of because it is designed for versatility and adaptation. Default thinking from an engineering perspective is that this redundancy is not optimized, or waste. It results in too much material. But if you think systemically and radically about the bigger picture, you can understand that redundancy actually enables difference, flexibility, circularity and therefore a more holistic way of thinking about architecture. This requires some imagination, but also an ability to step outside of your own understanding and to extend your knowledge of what should be to what other people think should be.

MG: People also want to take the path of least resistance. If you can offer something that is newer, and maybe a little slightly riskier, but easier—something that's easy for them to say 'yes' and to justify to their boss—it's almost like hiding the vegetables inside the dessert. If you package it the right way, as Molly is saying, you can start to initiate change. Getting that first person to say 'yes' unlocks so many doors for the next person to say yes, too.

GC: Speaking of industry, in this space, there's a very thin margin for failure. When we're in school, on the other hand, or in the lab, there are a lot of exciting things that come out of failure and experimentation. In Silicon Valley, there's this sort of move fast and break things idea. Maybe that's one perspective on failure. But I think maybe we can have other perspectives on failure and experimentation. How do you approach experimentation and non-success?

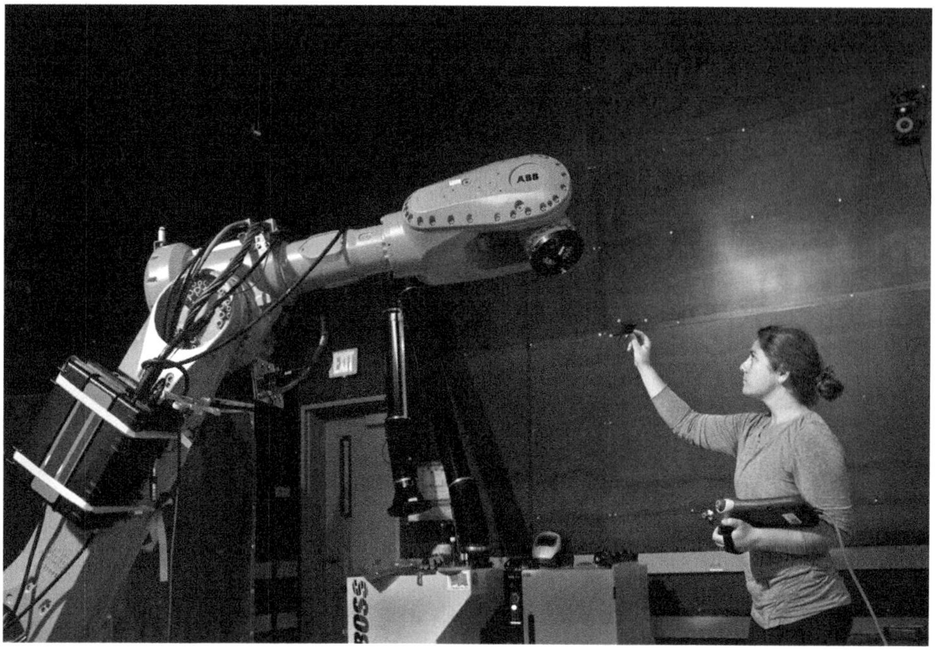

Figure 18.11 Madeline Gannon: "robot whisperer".

MC: My strategy right now with failure is very much the Silicon Valley mindset, which is to do it fast and see what happens, but, importantly, to do it with care. We are trying to crack open that little space where the focus is not how much profit we're making a month, or how many projects we're doing, but rather producing thoughtful and systematic ways of working and creating a culture of care around innovation and quick iteration. We are adamant that every iteration is a point of learning and thus allows us to improve. That's the way that nature works. Every version is the better version, most of the time. Unfortunately, this notion of care is not really baked into industry and practice. Care is not really a focus in school, nor in the current ways we work with each other. Industry is still all about bottom lines, by-the-numbers cost-cutting, etc. But if we can introduce a culture of care into how we work, it can help us understand the benefit of failure and how to make good things efficiently, and better, and most importantly, make sure that we don't do harm in the process (Figure 18.11)

BF: I agree with what Molly's saying. It's easier to experiment in academia. There's less risk and pressure to have a return on investment. Having said that, I have also seen a hesitancy to experiment in school at times. Sometimes there's this feeling that digital fabrication is too difficult or that it will take too long to learn the tool. In school, we need to convince students and colleagues that explorations might not return something tangibly valuable. A failed experiment can change how you value something. This is what I'm interested in. I don't think there is any failure in experimentation because an experiment is just a way to answer a question. No matter what the result, you have some answer to that initial question. Now, the smarter you ask those initial questions, the smarter your results will be. In my practice, I do many types of experiments: trying different tools, playing with different materials, using a wide range of technologies. I have a running wish-list of sensors that I want to experiment with and new programming languages I want to try. This means investing time in experimentation. It might take me six months or a year to get comfortable with a tool. The first 30 days are full of "failure" but after you start to see more positive results. My practice is very open to new directions because I think we always need to reinvent ourselves. If you stick to one way of doing things too long the work becomes less relevant. We have to keep experimenting and "failing" to stay fresh.

MG: I don't think of failure in the traditional sense. When you work in research, you work through experimentation: if you're not failing, then you're not exploring hard enough. My dad had a saying that has always been a helpful mantra for me, he said: *This is all just practice for the future.* That tends to take the pressure off whatever I'm doing. That's why we call it practice; it's a process that you invest in, and it grows and evolves with you. My sense of failure is when I'm in an R&D phase, and the results are just mediocre—or if it's something that someone else has done (or could do better!). That's my clue that I'm not asking the right questions or not looking far enough into the future. This kind of reflection and self-auditing happens in every project. I've gotten to a level of proficiency now where there are not many technical limitations to what I can do. Now the bottleneck is my imagination—it has been my biggest constraint, especially over the last five years. This year, I'm investing time into daydreaming and becoming more critical. This is a constant process of trying to be the best version of yourself and offer the best of what you can to the world.

GC: Do you see any kind of pressure to innovate in the realms that you are all working in?

MG: Yes, that's 100% the job! We must make new knowledge, new processes, and new things that don't exist in the world. We must make things that are better than what already exists. And at the same time realize that what exists is already pretty great.

GC: So how do you negotiate that pressure to innovate with your own individual values? Where might there be any points of friction?

MC: I strongly feel that my purpose in the world is to serve. I'm here to serve other human beings so that they can have a better world to live in. Innovation for me is a service. That's how I reconcile it. I am optimistic that the world can be a better place. I also have the privilege of being who I am, where I was born, etc. and these things give me resources to do the work. So there's no conflict.

BF: As three female practitioners, we strive to push the boundaries of innovation. Our mere presence in a traditionally male-dominated field makes our work inherently innovative. We consider ourselves disruptors, operating on the fringes of digital fabrication. Our sense of identity incorporates innovation, which often involves challenging the status quo and creating obstacles for conventional approaches.

MG: I also think a lot of the silos that we put ourselves in are unnecessary. I've basically made a career out of surfing in between all these different disciplines: taking a good idea that's normal over here, with a good idea that's normal over there, and smashing them together. Sometimes it's so obvious to connect these dots. But it's also frustrating because, for example, keeping the roboticists over here and the fashion designers over there only ends up limiting the potential knowledge that can be produced. There could be an interesting concept in Human-Computer Interaction and a related one in textile design, but because disciplinary walls exist, the two ideas may never meet. We should stop siloing ourselves.

MC: These conversations shouldn't be siloed. The more people we include the better, and I hope my work and practice is a good vehicle for enabling that.

BF: Maybe this goes back to the historical differences between the natural sciences and the social sciences. The former tends to focus on forces and systems, while the latter is more focused on social relations, power dynamics, etc. These have been historically separated from one another. But if we look at computational design and digital fabrication, we can see how both types of scientific work can come together. In digital fabrication, we can learn from natural systems, turn them into tangible objects, but also bring them to people and bring people into the process. We can have discussions about the social dynamics between the objects we make and people. We can push the field of digital fabrication to be a hybrid discipline that engages humanities research and very technical scientific research at the same time. The hard part is convincing practitioners to move in this direction. This is the trouble I'm interested in making.

GC: That seems to go along with the emergence of the DIY movement. The DIY movement and the "maker" movement made engineering and tinkering with electronics much more accessible. Now you can learn engineering concepts just by watching YouTube; you can learn how to make servo motors work and you can learn how to wire circuits from YouTube videos. But it's also a double-edge sword because on the one hand, more people have access to these tools and knowledge, which sound great, but on the other hand it demands the production of more materials and cheaper materials, which end up producing more junk and e-waste. What positions do you take on DIY culture.

MG: I'm a direct benefactor of the DIY movement. Around 2010, architecture education started to shift in response to the Maker Movement. There was a lot more emphasis on computation and making things with new technology. The ability to learn how to program or get technical answers online was quite transformative. In general, I believe that more access to knowledge is always better. The more people that have

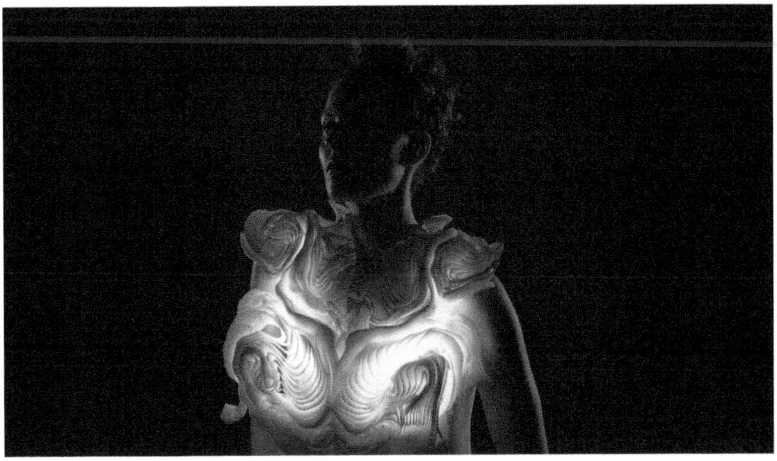

Figure 18.12 Bodyscape, Behnaz Farahi, 2017.

access to more knowledge, the more evenly distributed opportunities can be. I haven't studied the second- or third-order effects of this distribution of knowledge, but I really appreciate my experience during that peak Maker Movement moment. I was able to benefit from knowledge that was readily available online and not directly available in university courses in my field of study. I took my skills as a designer and augmented them with human-computer interaction concepts and engineering techniques to make new things. Then I became part of the creative coding community and ten years later, I still feel deeply attached to this group of people. I'm very grateful to have found that community, and to be able to give back to it.

MC: I'm a beneficiary of that as well. I've engaged that very question around wasting resources and how to set up frameworks for sharing physical products. For example, in one of the projects that we did in 2020, we built a prototype, and it has had many lives since then. It has been taken apart and reassembled. Pieces of it have gone out into the community; pieces of it have come back and other pieces of it have been turned into all kinds of strange and wonderful things. What we've found is that there has been a huge amount of ownership that people have taken over these parts that have gone out into their community, and that there is a network of relationships that have emerged to maintain and care for those parts. That was something that we didn't anticipate. I was under a lot of pressure to get the blocks out of their storage space. We decided to disassemble most of the prototype, and we didn't want it to be thrown away. So, we decided to have it be dispersed, reused, adapted, maintained, and cared for in its next lives. This is much more accessible now than ever before. It will also become more accessible as time goes on and will become more necessary as well as our resources continue to dwindle (Figure 18.12).

BF: It's also important to look at the different platforms that allow for this DIY knowledge sharing. Instructables, for example, has become quite mainstream and it's generated a worldwide community. Another example is the Processing community. These groups came together to help one another make interesting things. They are not competitive. There is less emphasis on ownership of ideas and more about helping each other achieve certain goals. Maybe I'm a bit too optimistic about it, but I really think that the power of these communities is incredible. It makes me wonder how important it is to talk about authorship. Especially with the increased use of AI generators and similar tools. How can we characterize crowd-sourced methods and human-machine collaborations? I think authorship in the traditional sense of owning a style or a technique is outdated. This kind of authorship is not really possible anymore because all of this knowledge is collective. Authorship is now more fluid and less about individuals. Even the individuals we associate with certain projects or artifacts are already influenced and informed by other people's code and ideas. There's no point in maintaining this narrow view.

GC: I'm a big fan of the creative coding community. I sort of slowly became a part of it over the last couple of years just through meeting different digital artists and getting to know them. The P5JS initiative that teaches JavaScript from a creative point of view is kind of amazing, because it has normalized this culture of tinkering and learning these tools as a community. Something that I found fascinating was the way they are introducing an aspect of care into learning programming. They have people invested in producing error messages that are not hostile or accusatory but point you in the right direction. They are also working on translating their documentation into other languages to make the library more accessible. These topics are not necessarily related to processes of making, but they play a huge role in designing and working with machines. Do you see any opportunities for these types of endeavors in digital fabrication? How do documentation and social interactions affect the culture of the fabrication lab, for example?

MC: Care is at the center of it. It is the point from which everything else radiates. Even the act of making is a form of care. No matter if you're doing it with a chisel or with code, you care about the process and what the outcome is going to be. A lot of the work we do at AUAR is about enhancing this dimension of the process. Since we are committed to giving people access to these tools, documentation and feedback are undeniably important. We have developed a system for the tool development that we're doing, where we ask for feedback much earlier in the process than you typically would in a typical software development project. In order for people to understand what they're looking at, we have to document it differently. We can't just show people what we're working on as a work in progress in its raw form as this would be too high a threshold for enduring engagement with them. We must package it and communicate it in a way that's meaningful for them so they can understand and digest it. That takes a lot of time and a lot of energy, but the benefit of that is that you get meaningful feedback. Too often consultation happens at a very late stage in the project and is not very meaningful at all. We use consultation, or our form of it, every step of the way. That transforms the way that people engage and take ownership of certain ideas. It also transforms the way that you think about your interactions with them, and where you can add value and meaning.

MG: Documentation in particular is such a labor of love. We document things for our future selves, but also so we know how to replicate something. This replicability is important in academia. For example, when you write a paper, the end goal is to transfer that knowledge from inside your head into someone else's head. Unfortunately, I don't think that mission is held as a standard in architecture. Conference papers tend to be summaries of what someone did and how interesting it looks. That's definitely a place for improvement. We need to incentivize learning more than the outputs. It would be great to measure the impact of a project by asking how many people have been able to replicate and riff off the project successfully, for instance. This would benefit the community more.

MC: This is the problem with the idea that an architect is a master. In the 20th-century model of the architect, novelty, and representation, the individual accomplishment is the thing that's most valued, or the singular man is the thing that's most valued. But in reality, we're here to serve the built environment and the people who live in it.

MG: Documentation can also be a huge pain. You can make a project open source, but if it's not documented well, then no one can use it. Open-source projects are often cumbersome, but when the project is useful enough, the community accepts it, and it can take on a life of its own. But it takes work and care to accomplish.

GC: Where do you see the digital fabrication and computational design heading? Homing the Machine is mostly about digital fabrication, but it's also inextricably tied to computational design. Do you have ideas about how the field might start to evolve or shift? Given this conversation, there are a lot of different ways that it can move forward. What are what are your visions for it?

MC: Generally speaking, digital fabrication is split into two worlds. There's the industry world and then there's the tinkerer world, which also includes academia. One of the things that is possible with digital fabrication is to think of it as a platform. That's the promise of the Fab Lab network that never came to be. That's what I would like to see. In a way, that's our business model for the company because that's where the sweet spot is. It's where we can retain knowledge, collaborate with people, and scale across industries. As a platform, we can have new kinds of conversations and enable opportunities that weren't there before.

BF: I was a member of the ACADIA (the Association for Computer Aided Design in Architecture) board of directors for many years. What has been happening in ACADIA, over the last few years from 2020, is the introduction of a critical edge into this technologically oriented community. Typically, the audience for this has been preoccupied with novel innovations and the newest advancements in computational design, robotics, and digital fabrication. But starting in 2020, we've started to focus on different ideas. In 2021, we organized the conference around 'critical computation' so we could bring in voices from other disciplines. We brought in people like sociologist Amelia Jones to address critical issues, including bias and critical thinking in design. I hope to see an integration of cultural and political issues into this field. We can begin to see how these technologies are not ends in themselves, but how they might help address larger issues. This is not only necessary but also urgent.

Index

Note: **Bold** page numbers refer to tables; *italic* page numbers refer to figures and page numbers followed by "n" denote endnotes.